Praise for Keith Stewart's

IT'S A LONG ROAD TO A TOMATO

"Keith Stewart's essays afford a fine way 'in' to the compelling realities of life on a small organic farm in the twenty-first century. His writing is precise and evocative—immediacy bound with a strong meditative underpinning that is an enduring pleasure to read. Like all really good writing, it illuminates a great deal more than the subject at hand."

    —SALLY SCHNEIDER, syndicated columnist and author of
     A New Way to Cook and *The Improvisational Cook*

"Keith's writing reads with the force and love of nature's elements—strong, refreshing, beautiful, and true. It's as fresh as his delicious carrots, and as poignant as his incomparable garlic!"

    —LESLIE MCEACHERN, owner of Angelica Kitchen in
    New York City and author of *The Angelica Home Kitchen*

"Keith Stewart has been providing New Yorkers with magnificent vegetables for two decades. Now, as if to prove he can do anything, he provides all Americans with a compelling story about his own approach to farming. And at precisely the right moment, just as millions of people across the country are rediscovering the pleasure, and the importance, of eating close to home."

    —BILL MCKIBBEN, author of *The End of Nature*,
     Deep Economy, and *Eaarth*

THE EXPERIMENT

BECAUSE EVERY BOOK IS A TEST OF NEW IDEAS

"Ever dreamed of living on a farm or growing your own food? Here's the clearest picture of what farm life really looks like. The romance of a pastoral life isn't shattered by Stewart's depiction of the gritty reality of farm life. They coexist, side by side, mirroring Stewart's organic and integrated approach to farming. Stewart's book is a gift to cooks. Now, each time I cook with food from a farmer I know, I have a deeper and clearer idea of what really goes into growing healthy and delicious food and why our farmers are heroes."

— PETER HOFFMAN, chef/owner of Savoy Restaurant, New York City

"To combat urban crowding, copies of *It's a Long Road to a Tomato* should be airlifted into major cities. The captivating charm of organic farming, so deliciously described in Keith Stewart's essays, would surely have hordes of city dwellers packing their bags. Stewart's stories transport me into the precious and full life of an organic farmer. I more than appreciate it; I now feel part of it."

—JEFFREY M. SMITH, author of *Seeds of Deception*

"Keith Stewart opens this engaging book by transforming himself abruptly from midlife executive into novice organic farmer. The twenty years that follow on an upstate New York farm are sampled here in true-life tales that—without denying the sometimes harsh realities of the small producer's life—leave the reader in no doubt of the joys that keep this small farmer on the land."

—JOAN DYE GUSSOW, author of *This Organic Life*

◆

"Here is a book from which a beginning farmer can draw not only inspiration, but volumes of practical knowledge and methods as well. Yet at the same time it is a book that us old hands can find virtue, honesty, and validation through . . . . And it is a book so carefully written and artistically illustrated that, I believe, anyone who picks it up will be drawn into its beauty and depth."

—*The Natural Farmer*

IT'S a LONG ROAD to a TOMATO

Tales of an Organic Farmer
Who Quit the Big City for the
(Not So) Simple Life

SECOND EDITION

KEITH STEWART

Illustrations by Flavia Bacarella
Foreword by Deborah Madison

THE EXPERIMENT

NEW YORK

It's a Long Road to a Tomato:
Tales of an Organic Farmer Who Quit the Big City
for the (Not So) Simple Life
Second Edition

The Experiment, LLC
260 Fifth Avenue
New York, NY 10001-6425
www.theexperimentpublishing.com

Library of Congress Control Number: 2010924689
ISBN 978-1-61519-023-2

Cover design by Michael Fusco | michaelfuscodesign.com
Cover illustration by Flavia Bacarella
Author photograph by Fran Collin
Design by Pauline Neuwirth, Neuwirth & Associates, Inc.

Manufactured in the United States of America

Second edition first published September 2010
Published simultaneously in Canada

10 9 8 7 6 5 4 3 2 1

To the whole of life,
in all its myriad forms.
Even the rocks
hold themselves together.

Good earth mother
Mountain and river
Taker and giver
In this hour of madness
I come to you
With blood on my knees

KS

CONTENTS

FOREWORD BY DEBORAH MADISON

IT'S a LONG ROAD to a TOMATO:
furthering the conversation

When it first came out, I thoroughly enjoyed Keith Stewart's tale of his farm, *It's a Long Road to a Tomato*. Reading it today with its new chapters and preface has proved just as absorbing as it was in 2006. Keith has a wicked wry humor, for one, but more importantly, *It's a Long Road to a Tomato* generously invites the reader into the complex world of this particular farmer and his farm, Keith's Farm. Anyone who has been lured by the succinct beauty of the small farm—as I have been on a regular basis—and anyone who has, more wisely, wondered what it takes to actually grow a serious amount food will get to see it all in this book. And we see it not just from the point of view of the farmer in charge, but also through the experiences of the interns who learn from Keith Stewart each summer and other integral members of the farm cast: the dog who showed up and stayed, the gift of a rooster that encouraged the idea of having chickens, the rescue of the hapless calf, the perusal of a beaver (before breakfast), and the antics and delights of birds and beasts of all kinds. We also see the life of the farm through Flavia Bacarella's beautiful woodcuts,

Above all, it's
the connection
to those who
grow, produce,
and otherwise
fashion the
food we
consume and
deeply enjoy
that matters.

which so strongly depict those moments that seduce, delight, and confirm the reason for our passion for farms.

But we learn about the other side, too—how easily loss is incurred when a truck filled with market produce breaks down, how seriously things can go awry, the mountain of paperwork involved in being a certified organic farmer, the difficulty of drought followed by too many downpours the next summer, and the vast amount of planning that goes into making twelve acres productive. And as we learn why farmers think and talk about weather so much, we see more clearly what it's like to work in intense heat or freezing cold, or to go days on end with not enough sleep.

◆

I met Keith Stewart at a book fair in Vermont where we were on a panel together. Just before we spoke, a group of us had been tasting some local heirloom apples and Vermont cheeses. I broke a tooth on a terrific apple, a Holstein, whose startling flavor so surpassed all the others that the costly dental work to come seemed a small price to pay for such a memorable fruit. And it's usually the case that produce from the farm stand or farmers' market, even when it costs more than food found elsewhere, is worth the extra money spent for the experience of its incomparable freshness, which I believe has its own sweet flavor. This is only one reason why it's easy to be an advocate for farmers' markets, CSAs, home gardens, and any other means of delivering food that is well grown and fresh from the earth directly to the table.

Of course, I'm hardly alone in this: A veritable band of cooks and writers has been pushing the idea of real food and farmers' markets forward for some time now. And one of the benefits that's often mentioned is the opportunity for connection. With what? With many things, but among them connection with others in our communities, the feel of air on our skin, or the season, as well as with concepts like foodsheds and watersheds. But above all, it's the connection to those who grow, produce, and otherwise fashion the food we consume and deeply enjoy that matters.

While farm foods have long been celebrated and elevated by chefs and food writers, farmers, oddly, have long been invisible and inaudible behind their beautiful bunches of chard and bouquets of radishes. Those of the kitchen realm were the primary voices not only praising produce, but also speaking to subjects that are more the everyday territory of farmers: the growing of organic food, the value of heirloom varieties and of saving seeds, the methods of chickens done right, the workings of integrated pest management, the value of harvesting food at the most opportune moment, and a great many other aspects of farm life. I myself was one of those who walked out of the kitchen and into the field on a regular basis and came back eager to report on what I had learned. But there came a point when I felt uneasy speaking on behalf of those actually growing the vegetables I loved. I wanted to know what *they* would say if they were telling their own stories. Wouldn't a farmer's words have more gravitas, more authenticity, and wouldn't they reveal worlds we hadn't imagined, but had only simplified and glorified?

◆

You might think that doing repetitive agricultural tasks would go hand in hand with the craft of writing. Surely the work would present ample opportunities for mulling over ideas, turning a phrase, or finding that right word. However, after reading *It's a Long Road to a Tomato*, one might well wonder when Keith managed to write. The work of the farmer, we learn, is truly without end, or at least its end is a very exhausting one. But however it happened, Keith Stewart has told his story.

It's a Long Road to a Tomato reveals those moments of rare beauty and insight that occur in the life of a farm. It also shows that farming is not for the fainthearted. One of the new chapters, in particular, brings this home—the chapter that deals with the 2009 summer of rain in the Northeast. A lot of us read about the tomato blight as it was unfolding, and many more experienced it from the point of view of the disappointed consumer:

"What, no tomatoes? Again?"

That was a sobering summer.

But who heard about the heartbreaking moment when hundreds of tomato plants were thrown into a hole and covered with earth, or when other crops were pulled and destroyed before harvest—necessary practices that also ensured a financially devastating season? Getting the bad news from one who experienced it shows us how delicate our fields of food really are, and how that connection of eater to farmer might easily become frayed without understanding and compassion.

It's a Long Road to a Tomato is a heartfelt chronicle, sobering and amusing by turns. Although focused on the particular, it transcends Keith's Farm and illuminates exactly what it is that we are putting on our plates, whether we shop at Keith Stewart's stand in the Union Square Greenmarket or at a farmers' market elsewhere. It's a delicious read—but what makes it an important one is that it has so enriched the ongoing conversation about food. How good it is to hear the voice of the farmer. How wonderful to learn from him what happens on his patch of earth day after day and season by season. And how essential it is to know what it takes to get that tomato all the way from newly tilled field to you. It makes us more food-literate beings, and possibly more understanding ones too. This is the voice that has been missing, and it's good to hear it.

DEBORAH MADISON *is the award-winning author of eleven cookbooks, including* Vegetarian Cooking for Everyone *and* Local Flavors: Cooking and Eating from America's Farmers' Markets. *She has been active for many years on the issues of biodiversity, seasonal and local eating, farmers' markets, and small and mid-scale farming. She writes for both culinary and garden magazines, and while she has never attempted to farm, she does manage to grow some vegetables at her home in Galisteo, New Mexico.*

PREFACE to the SECOND EDITION

In my life, 1986 was a year of change: I married my long-time girl-friend, Flavia Bacarella, said goodbye to the big city, and set out to become a farmer. This book is the story of what followed. It has not been all roses. But it has mostly been a good story, or, I should say, the right story. Even on days when the sun was not shining, I knew that being a farmer was the best choice for me. But what is a little surprising is that I really had no inkling of this until I had spent nearly forty years of my life doing other things.

Most of the chapters in this book are free-standing essays written over the past twelve years, though some recount stories that reach back to earlier times. They are not always presented in chronological order. Some trace my often-faltering development as an organic farmer; some try to give expression to what I feel is unsound in the way we feed ourselves and treat our planet; others recount the more memorable, at least to me, experiences that twenty-three years of farming have provided.

The book was first published in 2006. In the four years since then, life, of course, has not stood still. The farm has continued to operate and flourish: More young workers have labored in the fields; more singular cats, dogs, and chickens have shared the land with us; more crops gone to market; more generations of barn swallows come and gone; more geese passed overhead. And there have been more farming stories to tell, some of which have found their way into this new, expanded, and updated edition of *It's a Long Road to a Tomato*.

"Breakdown: Perils of the Truck-Farming Life" describes what it feels like when your truck fails on the way home from market, after a very long day. "A Beaver before Breakfast" is the tale of an early-morning encounter with an unusual and impressive visitor. "About Seeds" is an attempt to elucidate the distinctions between different kinds of seeds—be they heirloom, hybrid, open-pollinated, or genetically modified.

If I ever had any doubts about the title chosen for my book, they were dispelled in 2009. That was a very bad year for tomatoes. Along with many other growers up and down the east coast, we suffered the ravages of late blight, a disease that can wipe out entire fields of tomatoes and potatoes within days. And it did wipe out almost all of our tomatoes, so that the road to this especially valued vegetable (or fruit, if you prefer) became even longer. Luckily, the potatoes fared better. One of the chapters in this new edition tells of our struggles with late blight, the disease that sent a million Irish peasants to their graves.

But 2009 also had its high points. Michelle Obama had the imagination and the mettle to plant an organic vegetable garden on the White House lawn, for which she was severely criticized by the Mid-America Crop Life Association and other groups that promote the use of agricultural chemicals—to me, this seems a bit like a heavy smoker being infuriated by others who have made the choice not to smoke. And Tom Vilsack, our nation's secretary of agriculture, took a jackhammer to the parking lot in front of the Department of Agriculture, as a first step toward installing an organic garden there too. I took these to be positive developments. Vilsack said that he would like to eat more vegetables and perhaps thereby live a longer and healthier life so that he might enjoy his grandchildren—something his own parents did not live long enough to do. That makes sense to me.

But, most significantly, in 2009, even with the economy in dire straits, unemployment at record levels, and a sense of disquiet across the land, it was clear to me that a growing number of my fellow countrymen and women were not about to turn their backs on fresh, local food that tastes good and is grown or raised by people they can actually talk to. It is heartening to know this. Like Tom Vilsack, more and more Americans understand that there is a direct connection between what they eat and the way they feel, and perhaps even a connection between what they eat and their actual longevity. As we confront our broken healthcare system and epidemic levels of obesity, heart disease, and diabetes, should not a consideration of the food we eat—the very fuel that keeps our bodies running—be front and center in any discussion? Certainly, we have many miles to go and powerful interests standing in the way, but I am confident that the public's enthusiasm for local food, produced in a sustainable manner, will only increase. And, as it does, it is my hope that we will move toward a healthier, happier, and more earth-friendly tomorrow. Of course, in order for this to happen, we will need new generations of small, diversified farmers scattered across this rich and sparkling land. If this book persuades just one man or woman to take up the agrarian life, it will have done its job.

The last chapter of this new edition of *Tomato* describes a major change in status of the farm on which my wife and I live. In 2007, a conservation easement was placed on the land which protects it from future development and ensures that it will remain open space "in perpetuity." For both of us, this was an important achievement that we had been working toward for several years. The idea of a housing development or strip mall, even in the distant future, on the land we have become so intimately involved with was anathema to us. When it comes time to move on, we will know that this old farm, with its glistening ponds, its many and diverse inhabitants, its fertile fields and rocky woods, will keep on going.

We will need new generations of small, diversified farmers scattered across this rich and sparkling land.

KEITH STEWART
February 2010

The Farmer and His Dog

A CHANGE of LIFE: on becoming a farmer

Twenty-four years ago, a little past the age of forty, I was living in a small apartment in New York City, working as a project manager for a consulting firm, wearing a jacket and tie to the office every day. It didn't feel good. I had never aspired to be a member of the corporate world, but somehow that's where I had ended up. I had little affection for the work I was doing and seldom experienced any feelings of pride or fulfillment. Rather, I felt like an impostor, obliged to feign interest and enthusiasm much of the time.

I also felt that time was running out, that I was moving rapidly into middle age, that my life was getting used up with not much to show for it. Both my body and my disposition suffered from chronic low back pain, and the fitness of my youth seemed long gone. Colds and flu and other ailments were common occurrences in my life. Most mornings, as I got nearer to the office, a heaviness would settle into the pit of my stomach. Finally, there I was. I'd be going up in the elevator, but my spirits were coming down as I readied myself for the hours that lay ahead. There was nothing wrong with the work I was doing. But it wasn't right for me.

Today I am a farmer, a grower of organic vegetables and herbs, and I can honestly say that I am a happier man. True, I work more hours, have no company retirement plan or paid vacation, and have more things to worry about. But I have less back trouble than I used to; I rarely catch a cold; and I have almost forgotten what it's like to be down with the flu. I enjoy good food and a midday nap and I sleep soundly at night. I've lost weight and put on some muscle around the shoulders. The shirts and jackets of my earlier life soon became too small for me and have long since gone to the Salvation Army. My life now is more full, more varied, and more interesting. Often it is more demanding and exhausting, but it is always more real. I've never for one moment thought of going back to the old days.

It started as a yearning: to live on a piece of land, closer to nature; to work outside with my body as well as my brain; to leave behind the world of briefcases, computers, corporate clients, and non-opening windows. I knew next to nothing about growing vegetables, but I had always had a love of land and wild places. When I was younger I had thought of being a forest ranger, a game warden, a wilderness guide, but by age forty my needs and desires were more modest. Ten or twenty acres of good land—a small farm, a place with lots of life on it, a place to put down roots and live more in accord with my environmental inclinations—seemed like just the ticket.

My then girlfriend and soon-to-be wife, Flavia Bacarella, and I began making weekend excursions from New York City to Orange County in upstate New York to visit real estate brokers. It was quite

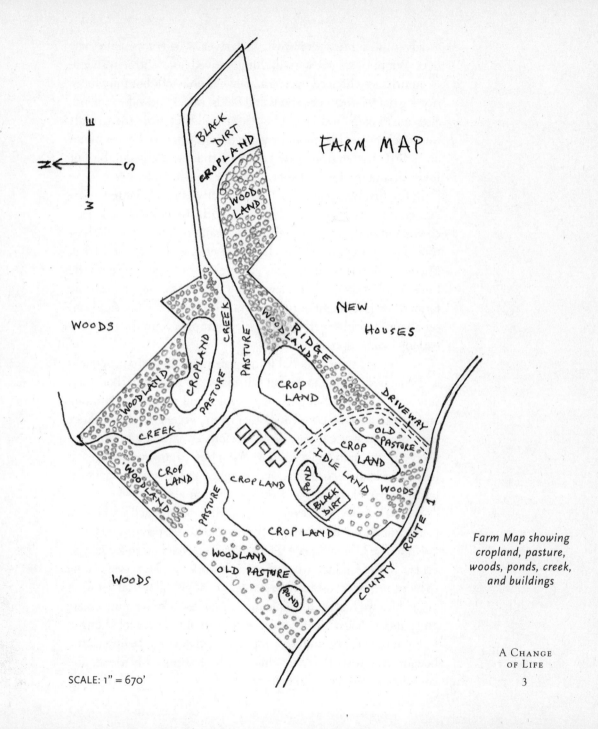

FARM MAP

BLACK DIRT CROPLAND

WOOD LAND

WOODS

NEW HOUSES

CREEK

WOODLAND

CROPLAND

PASTURE

PASTURE

RIDGE

WOODLAND

CROP LAND

DRIVEWAY

CREEK

OLD PASTURE

WOODS

CROP LAND

WOODLAND

CROP LAND

PASTURE

CROPLAND

IDLE LAND

POND

BLACK DIRT

CROP LAND

WOODLAND OLD PASTURE

POND

WOODS

COUNTY ROUTE 1

Farm Map showing cropland, pasture, woods, ponds, creek, and buildings

SCALE: 1" = 670'

an adventure for two people who had never owned property in their lives. But once we got started, things moved fast. Within a couple of months we chanced upon a somewhat unkempt but fully functioning dairy farm, with woods and fields, ridges and vales, a pond, a stream, a barn, and an old house set well back from the road. It was much bigger and more expensive than what we had in mind, and a little farther from New York City than we wanted to be, but it was definitely the right spot. I knew it immediately.

In the first half year or so, I kept my city job and planted a garden behind the house on weekends. With the help of some aged cow manure, things grew with wild abandon. We had bushels of tomatoes and enormous zucchini and peas and beans and basil. Encouraged by this early success, the next year I signed up with the Greenmarket program in New York City and became a full-time farmer. I paid a neighbor to plow and disk an acre of hay field and set about planting everything that caught my eye in the first seed catalog I came upon.

The second time around, I learned about weeds and woodchucks and what lack of rain can do. But still, many plants grew and bore fruit; tiny seedlings turned into heads of lettuce and escarole and mizuna and tatsoi and red mustard and other exotic vegetables that had not even been in my vocabulary the year before. I worked hard in the field every day, and at night I read books about how to grow vegetables and live on the land. I was inspired and energized and relatively undaunted by the inevitable failures and setbacks that came my way.

Each Thursday, whatever looked good enough to sell got picked, packed, and loaded onto the back of an old Dodge pickup and taken down to the Union Square Greenmarket in Manhattan for sale on Friday. In the summer, when she wasn't working, Flavia came along to help me. We set up a ten-by-ten-foot canopy and spread out what vegetables we had on a couple of card tables. We hung up a scale and put out a cigar box for the cash and an old ashtray for change.

From the very beginning, people seemed delighted to see us and bought what we had to offer almost unhesitatingly. My friends and co-workers in the city had all thought I was a little crazy, that my "back to the land" urge would be short-lived. I was beginning to think that, contrary to popular wisdom, I might actually make a living as a farmer. (It didn't happen in the first year, or the second, but in the third year I turned a small profit and almost every year since then has been better than the one before.)

In the third year I took on my first assistant, Mitch, an eighteen-year-old vegetarian college student, and started going to a second market, at the World Trade Center. Mitch and I got along well and work got done much faster with another pair of hands. When he went back to school in August, a couple who had just returned from a Peace Corps stint in Ecuador came to replace him. They also made a nice addition to the farm. I was on my way to becoming an agricultural employer.

There has been much to learn. But when your heart agrees with what you are doing, the learning is easier and more fun. Now, more than twenty years on, I am firmly established as a small farmer. I make a moderate living and have been able to reinvest some profits back into the farm.

"If you want to succeed as a small farmer," one farmer I know once said, "you better do it as though your life depends on it." He may be right. Running a small, diversified, organic farm in today's environment of industrial agriculture, chemicals, and cheap food is a bit like swimming against the current. It taxes every muscle and sinew in your body, and a few in your brain as well.

But for those who are drawn to it, farming offers a different kind of life and an assortment of rewards and satisfactions not readily found in other types of work. It is performed in the outdoors, in the realm of the sun, the wind, and the rain. It is varied and vigorous work. The farmer is his or her own boss. We make decisions for better or worse and move on. We deal with tangible, living things. We see the fruits of our labor and the results of our neglect. We are on good terms with the natural world, or we should be, and we inhabit it in a practical, down-to-earth way.

Across the United States, the prognosis for small farmers is not good. Many are struggling to make a living and stay in business.

Others feel marginalized and unappreciated. Each year, too many give up their land and look for other work. But in the midst of this grim picture, there are some impressively bright spots. While many rural communities are in decline, paradoxically, in and around large urban areas there is a resurgence of interest in small farms and the fresh, wholesome food they provide.

In the Hudson Valley and New York City, it has become quite fashionable to be a small farmer, especially an organic one, selling directly to the public. I did not anticipate this when I started out in 1987. But I'm not complaining. We get a lot of good press, a lot of good feedback. I've done many other kinds of work in my life, but none where I felt as appreciated and needed.

I hope this sentiment will grow and spread. And I hope it will induce others—the young and not-so-young men and women who are looking for a change of life—to give farming a try. It's a challenge, to be sure, and a bit of a gamble, too, but it might put some spunk and spirit into your life.

REGARDING CHICKENS and THEIR EGGS

I t was neither a chicken nor an egg that came first on our farm.

It was a rooster. He was large and white and he came in a burlap

sack. Don Lewis, my neighbor at the Greenmarket in Manhattan,

had brought him down as a special order for a Jamaican customer

who, presumably, had a culinary event in mind. But the fellow never

showed up, and at the end of the day Don was left holding the bag

with a live rooster and two more nights in New York City. Concerned

that his host, a late riser, might not appreciate being wakened by a

rooster in his downtown apartment, Don suggested that I take the bird home with me. At the time, I had no plans to introduce poultry to our farm, but the notion that fate or chance or just dumb luck might so intervene to offer this unfortunate bird a reprieve had some appeal to me. Out of a blend of curiosity and compassion, I accepted.

I peeked in and, to my delight, discovered three beautiful pale blue-green eggs.

Back at the farm that night the rooster tumbled from the sack in a somewhat dazed condition. But he quickly found his footing, assessed his new surroundings, and apparently determined that he had arrived in a foreign and possibly hostile environment. He proceeded to run up our driveway at great speed, emitting a series of raucous shrieks and using both legs and wings to transport himself. Under the light of a full moon, perhaps unwisely, I gave chase. This only spurred him on to run faster and shriek louder.

Eventually, he got airborne long enough to propel himself onto the branch of a tree, just out of my reach. I yelled at him and shook the tree a few times, but he wouldn't budge. Tired from a long day at the market, I accepted that he had got the better of me and turned back toward the house.

The next morning, I spotted the rooster behind the barn scratching about in a pile of aged manure. I let him be and went about my chores. That night I noticed him perched on a rafter in my tractor shed. A few days went by. I acquired some cracked corn and stale bread and threw a few handfuls in his direction. These offerings made all the difference: I was his new benefactor and he knew it.

On my next trip to the Greenmarket I mentioned to Don that the rooster had settled in quite nicely but that perhaps his life would be more complete if there were a hen or two in it. Don agreed; the following week he brought me a beautiful, auburn-red bird with a calm disposition and a rather regal bearing. She would remain with us for several years and become the matriarch of a much larger flock. In time, we named her Henrietta.

Henrietta and the white rooster seemed quite compatible. They spent their days in the yard, seldom out of each other's sight,

Henrietta

enjoying a diet of plants and insects and a daily ration of corn. At night they retired together to the rafters of the shed from where they decorated the hood of my tractor with large droppings, colored according to their preferred food of the day (mulberries being easiest to detect).

Early one morning I noticed light reflecting from an oval object inside an unoccupied doghouse near the tractor shed. I peeked in and, to my delight, discovered three beautiful pale blue-green eggs. I called to my wife, Flavia, to come and take a look. She did so and was suitably impressed. We decided to have an omelet for breakfast.

Henrietta, it turned out, was an Araucana, a South American bird sometimes known as the "Easter Egg Chicken" for the pastel blue and green eggs they lay. I am tempted to suggest that she was the inspiration behind Martha Stewart's "Araucana Colors" (a line of towels and bath ware offered to the public some years ago), but this is unlikely since Martha Stewart never met Henrietta in person, and she and I, though we share a common name, are quite unrelated.

Within a few short weeks, misfortune befell the white rooster—in what form we will never know. One day he simply disappeared and was not seen again, though we noted a telltale scattering of small white feathers along the driveway. Fate, it seemed, had intervened in his life once again—but this time she had expressed her darker side. I passed the news on to Don, who assured me that he could repair the situation. A week later he brought me another rooster, a striking multicolored bird, also of Araucana stock.

Henrietta, who appeared unfazed by the loss of her former mate, happily accepted the new one. Within days, she and he were seen engaged in brief amorous encounters, and not long after Henrietta must have decided it was time to start a family. She had wisely accumulated a clutch of seven or eight eggs in the doghouse and now proceeded to set on them in a detached, trancelike state for about three weeks. Each day I provided her with a handful of corn and some fresh water. She partook of these victuals but otherwise seemed oblivious to my presence. On a rainy Friday in August,

Henrietta's patience and dedication were rewarded with five tiny chicks. (Two of the eggs were pushed off into a corner of the dog-house, apparently discarded as unworthy. I cracked one of them open and was met with an extremely foul odor and the sudden urge to create as much distance between that egg and me as possible.)

The chicks spent most of their early days under Henrietta's wings, coming out now and then to make high-pitched peeping sounds and peck about in the manner of their doting mother. We provided some hay for bedding, some finely ground corn, and a tuna-fish can full of water, which the chicks frequently stepped into and Henrietta knocked over at least once a day. Within a few months the chicks grew to be almost full-size birds. Two, it turned out, were roosters. They quickly assumed subordinate status with regard to their father but occasionally engaged in aggressive behavior toward each other. Like awkward teenage boys, they ran about with frenetic energy, yet seemed unsure of their role in the world to which they were consigned. The other three chicks turned into hens. For a while they did everything together and it was hard to distinguish one from the other, then gradually they developed their own traits and personalities and degrees of dominance. The following spring, each began laying small pastel-colored eggs that we gathered and ate whenever we could find them. It seemed by then that we were well on our way to having a real flock of chickens.

Though still strictly amateurs in the art of chicken husbandry, we have come a long way since the white rooster and Henrietta's first brood. Today we have sixteen hens and just one rooster. He is adequate to service the flock and will not tolerate competition. A few years back Henrietta died, apparently while attempting to lay an egg. We buried her under a pile of dead branches and recognized her passing with a small ceremony.

Henrietta's progeny are still with us, along with some newcomers. They do not range as freely as she did; instead, they are contained within a large enclosure, partly for their own protection from predators (including our own dogs) but also for the protection of our green crops, many of which the hens are fond of eating.

Factory
chickens,
permanently
confined to
laying cages,
do not lead
well-rounded
lives.

In the long days of spring and summer, when the hens are laying at their peak, we get about fifty eggs a week. The older hens lay less frequently than the young ones (and possibly not at all), but we let them live out their lives. In late fall and winter, when the days are short, all the birds take a sabbatical and no eggs are laid.

Most of the eggs produced on the farm are consumed by me, my wife, and our crew of workers. Some are given to close friends and neighbors—the eggs are regarded as a gift worth waiting for. A friend who receives a dozen or so eggs a year recently asked, "What kind of chickens do you have? Their eggs are incredibly good." I told him I don't think the breed is relevant, beyond the obvious difference in shell color. "It's all in the diet," I said, "and perhaps the birds' ability to move about outside and behave like real chickens."

Our birds eat almost anything we give them, including all manner of kitchen leftovers that are no longer appetizing to humans— from spaghetti and potatoes to aged cottage cheese and sour cream. They love grapes and tomatoes and apples and most green vegetables. They eat vast quantities of market leftovers, including old mesclun salad, kale, collards, and entire heads of lettuce. They like the seeds inside summer and winter squash and will eat the flesh as well. They are broadly omnivorous and will gladly eat other kinds of flesh, too—even chicken, if it is offered to them. This I discovered inadvertently, after throwing them a pile of unsorted kitchen scraps.

We believe the eggs from our chickens are superior in every respect to store-bought eggs. The shells are hard and well formed (a handful of ground oyster shells tossed into their enclosure every other week ensures this), and the colors are irresistible.

But the real treasure is found inside: The yolks of these eggs are like little orange suns, firm and radiant. They separate easily from the perfectly intact whites. My wife, an artist, likes to describe the yolks as "cadmium yellow deep," after the color of a paint she uses. Our eggs are never pallid, watery, and lifeless, like so many supermarket eggs. Cracking them open is part of the joy of eating them.

Factory chickens, permanently confined to laying cages, do not lead well-rounded lives—they are strictly egg-producing machines.

Roosting Hens

But they barely get to see the fruits of their labor. Each time an egg is laid (about every twenty-six hours) it rolls down the sloping floor of the hens' cage and disappears onto a conveyor belt. These unfortunate birds are crowded by the hundreds of thousands, sometimes even millions, into agro-industrial facilities. They are housed four or more (sometimes as many as eight) to cages about the size of dresser drawers. Often they are barely able to turn around. Their beaks are cut off so they will not attack and possibly cannibalize each other in their boredom and frustration. They spend their waking hours under controlled, artificial light and heat so that they will lay year-round. When most of a hen's eggs are laid (there is a finite supply within each bird), she is of little value, except perhaps for cat food or soup—off with her head! The mere thought of eating the eggs of animals so cruelly exploited is distasteful.

Our hens lead very real lives by comparison. They have ample outdoor space and adequate protection, though on occasion a predator will penetrate their enclosure and take its toll. In the winter they live in the barn, where they congregate together to stay warm. They establish social hierarchies and interact with each other constantly, though often harshly. (If you've ever wondered where the term *pecking order* comes from, simply observe a flock of chickens for a couple of hours.)

Our hens live in the presence of a rooster who rouses them in the morning, alerts them to danger, calls them to their food, and fertilizes their eggs. If they are so inclined, the hens may sit on their eggs and hatch them into chicks, assuming the rooster has been doing his job. They enjoy a varied and healthy diet. They experience the vagaries of the weather and the changing seasons. In short, they are challenged to be what they are—chickens. And I believe you can see the difference, taste the difference, and perhaps even feel the difference, when you eat their eggs.

BUY IT at the FARMERS' MARKET

Picture this scene: a crisp, clear morning in early fall; a line of trucks and hastily erected canopies; men and women, their sleeves rolled up, unloading trays of bread, boxes of jams and cheeses, setting up displays of vegetables, fruits, flowers, even fish. Baskets overflowing with shining black eggplants, crates of red and yellow tomatoes, heaps of broccoli, cauliflower, green and golden squash, peppers of all colors and shapes, carrots that don't look like they all came out of a mold, potatoes with the dirt of the field still on them. Where else, but at a farmers' market, is nature's bounty so evident, so proudly on display?

Where else can you be seduced by the fragrance of fresh-cut basil, bunches of sage, rosemary, thyme, marjoram, dill, oregano, and herbs even more exotic than these? When did you last see shoppers in a supermarket bending over to smell the produce? At farmers' markets you see this all the time, because the scent of freshness is still there to be inhaled and enjoyed.

Where else can you find people engaged in commerce and so evidently enjoying it—the small talk, the jokes, the laughter, the exchange of recipes and advice, the feeling of community, the happy confluence of seller and buyer, grower and eater? Farmers' markets have come a long way since John McPhee wrote his vivid essay, "Giving Good Weight," on the early days of the Greenmarket in New York City, but his description of the flow of the crowd still holds true today:

> "There is a rhythm in the crowd, in the stopping, the selecting, the moving on—the time unconsciously budgeted to assess one farm against another, to convict a tomato, to choose a peach."

In 1970 there were a half-dozen or so farmers' markets in all of New York State. In 2009, there were well over 400, with approximately 2,500 farmers participating. Revenues from these markets are in the tens of millions of dollars annually. And this phenomenon is not confined to New York State. A survey conducted by the USDA in 2009 counted 5,274 farmers' markets nationwide, up from 2,863 in the year 2000—that's an increase of eighty-five percent over a nine-year period. Farmers' markets have made a comeback, and for good reason. Whether you're a farmer, a shopper, or just passing through, these days the market is a great place to be.

In many instances, farmers' markets have helped to draw people back into neglected downtown areas that have lost customers to malls and shopping centers on the outskirts of town. In the area in which we live, the cities of Middletown, Kingston, Newburgh, and Poughkeepsie are cases in point. Though sometimes initially resisted, farmers' markets soon become popular with local merchants, municipal authorities, and, of course, area residents.

Union Square Farmers' Market

Some years back, when a New York City Parks Department plan proposed trading a dozen or so farmers' spaces at the Union Square Greenmarket in Manhattan for two rows of elm trees and a walkway, the public protested in the form of more than nineteen thousand signatures and thousands of letters opposing the plan. And these are people who like trees! But they value the market even more, because they recognize the positive impact it has had on their lives. To them it is an oasis, not just a place to come for fresh, good-tasting food, but also a place to come for the replenishment of their spirits. Like the town square of earlier times, the market is a commercial and a social environment, a place where local residents come to mingle and chat, to enjoy one another's company and be reminded of their common humanity.

Farmers' markets have had a very positive impact on farmers' lives, too. Many of us would be in some other line of work today were it not for the resurgence of interest in locally grown food and the open market where consumers can buy direct from the source. It has become increasingly difficult for farmers here in the Northeast, with our short growing season and high production costs, to compete at the wholesale level with large growers in California, Florida, and beyond.

Today, the tomatoes in your local supermarket are likely to come from Mexico, the strawberries from Guatemala, the grapes from Chile—all places where labor costs are low and regulations to protect workers and the environment are few. The new "global economy" may be good for the agribusiness giants, but it can be deadly for small local growers.

Farmers' markets offer one of the few remaining economic niches where small farmers can still make a living. Being able to capture the retail food dollar makes all the difference. Plus, it's both enlivening and enlightening to leave the farm once or twice a week to interact with the people who are eating—and enjoying—what you grow.

People come to farmers' markets for food that is fresh, nutritious, in season, and good tasting. They are looking for alternatives

Farmers' markets offer one of the few remaining economic niches where small farmers can still make a living.

to the junk foods, fast foods, and highly processed foods that are so much a part of the American diet today. In a recent survey, New Jersey farmers' market customers said they had greatly increased their consumption of fresh fruits and vegetables since they began buying from farmers at open-air markets.

Many farmers' markets take an active role in promoting nutrition education and an interest in local fresh food. Chefs from nearby restaurants (many of them market customers themselves) are often invited to come to the market to prepare sample plates of their specialties, using local ingredients provided by the farmers. Customers taste the samples, meet the chefs, and may be inspired either to buy from the farmers or visit the chefs' restaurants—or possibly both.

Even the federal government has recognized the value of fresh produce purchased directly from farmers. Lower-income women, infants and children, and seniors, all of whom may be nutritionally at risk, can participate in the Farmers' Market Nutrition Program (FMNP), established by the USDA in 1992. Via coupons that can be redeemed only at farmers' markets, this program encourages recipients to purchase fresh, unprocessed fruits and vegetables. Today, the FMNP is a highly successful program with bipartisan support in Congress. It has boosted small farmers' earnings and introduced lower-income people to the rewards of eating fresh, local food.

There is another way in which farmers' markets help to feed the less fortunate among us: They provide a great opportunity for food-gleaning and food-recovery services that cater to the poor and the homeless. Nonprofit groups such as City Harvest in New York City, with its fleet of trucks and army of young volunteers, pick up bags of perishable food donated by farmers at the end of a market day. Annually, thousands of tons of food acquired in this manner are distributed to the poor.

If all your grocery money still goes into the supermarket cash register, it may be time you stepped out into fresher fields. Chances are, this spring there will be a farmers' market not far from where you live. Pay it a visit when the harvest is at its most bountiful. You'll almost certainly have fun, and you just might become a convert. You might find yourself joining the growing number of Americans who have discovered that the local farmers' market is just too good to pass up.

Sweet Dumpling
Squash

AN APPRENTICE
WORKFORCE

In my early years of selling at farmers' markets, there were two questions I got asked a lot: "Is she your daughter?" and "He's your son, isn't he? You look so alike." "No, no," I'd say, with a chuckle, "she's an employee" or "he works for me," sensing a flicker of disappointment in the eyes of the questioning customer. It seems that people like to think of the farm family including a few kids running around. And I can understand that.

As for those regular customers who might have wondered about the parentage of my workers but never ventured to ask, they must

Potato Diggers

know by now that I have no paternal involvement, after having seen so many new young faces at our stand from one year to the next. Either that, or they've come to the conclusion that I have a very large family. But the truth is, I do not have children, nor do I feel the poorer for it.

Each spring, my first and greatest challenge is to put together a crew of workers—usually six young men and women—who will live on the farm for up to eight or even nine months and spend their days planting, tending, weeding, mulching, watering, picking, and selling the vegetables and herbs we grow. Organic farming is very labor intensive. More than weeds, insects, disease, floods, drought, or hail, my crew's reliability, their willingness to work, and their ability to coexist amicably will determine whether or not we have a profitable year.

Most farmers in my area hire migrant workers, usually from Mexico, who are strong and accustomed to long hours of tedious labor. They work for the money and they know how to get a job done. We've looked into hiring migrants but have thus far declined.

Instead, we offer an "organic farm internship." We take young (and sometimes not so young), mostly well-educated, middle-class Americans. We give them housing and food that is grown on the farm, and we offer a stipend that is somewhat less than migrant workers receive. We provide on-the-job training that allows for plenty of questions and answers and discussion of organic farming techniques and philosophy. And we involve each person in a wide variety of tasks and responsibilities.

It's a good way to learn about our kind of farming. In fact, it's almost the only way to learn, since very few agricultural schools in the United States offer instruction in organic methods. The agri-business funding sources that agricultural schools often depend on are not interested in promoting organic or low-input farming—i.e., the type of farming that does not depend heavily on purchased inputs, most notably a host of expensive synthetic pesticides and fertilizers.

Most of those who come to our farm have college degrees and a range of employment opportunities. Not that this necessarily commends them to me. I have sometimes wondered if there might be an inverse relationship between an individual's level of academic qualification and his or her ability to actually do physical work. In my years as an agricultural employer, I've seen a large sample of educational accomplishment, going all the way up to the doctoral level. At this point, the evidence is inconclusive. The subject might, however, at some time in the future, merit a scholarly study-which, to be fair, would have to include me.

In any event, these young people, whether burdened with academic achievement or not, and frequently to the puzzlement and chagrin of their parents and teachers, actually *want* to work on a farm. They usually have environmentalist leanings. Frequently they are vegetarians. Invariably, they are disenchanted with the impersonal nature of our nation's corporate food system and its heavy reliance on chemicals and monoculture. They are not convinced that the way things are is the way they have to be.

Interns come to our place (and to hundreds of other small farms around the country) to find out what it's like to work under the

It's a good way to learn about our kind of farming. In fact, it's almost the only way to learn . . .

sun with their bodies as well as their minds, to find out if organic farming is a viable option or just a pipe dream, to discover if it might be right for them. If they stay to the end of the season (most do), they will have seen that it is possible to grow decent food and live in relative harmony with nature; they will have learned what it means to be tired at the end of a day's work on the land; and they will have tasted an old style of living that could become new again.

Here's what some of our interns over the years have had to say about their farming experiences and their reasons for giving farming a try:

෴ **Ione Lindroth** – I grew up surrounded by citrus monoculture, and for years I watched most of the hard labor being done by Mexican migrant workers. As a blond white female, to me it seemed farmwork to be something I would never experience, much less enjoy.

In college, I majored in soil science; after graduation I continued my research into alternative agriculture, especially soil building. I worked several jobs—from testing soil in laboratories to outdoor environmental education—but a deep, restless part of me wanted to work and live on an organic farm where cars and asphalt were not part of the daily routine.

Next thing I knew, I was weeding carrots at Keith's Farm, watching a great blue heron glide in for a landing. For seven months I didn't drive to work—I rolled out of bed and walked out to the land. Dress up for work today? Wear nylons? Put on deodorant? Comb my hair? *Nah!* Nature does not care what I look like.

Growing a hundred different vegetables and herbs is no small task, but it was rewarding when I saw the looks on the faces of the customers at the Greenmarket. But some customers were not so pleasant. Some saw the prices and pranced off mumbling under their breath. This revealed to me how the traditional cheap price of food in America has skewed the values we put on our nourishment and on the work of the farmer. Most people are so distant from the source of their food they forget how many hands toiled to get that onion on their hamburger.

Gandhi once said, "You must be the change you wish to see in the world." I like growing vegetables. I like not worrying about chemical residues in my food. I like growing soil, not just plants. This is the change I wish to be and see in the world.

❧ Sandra Cavalieri — It was fear that began to dominate the last days of my senior year at Barnard College: Where would I go? What could I do? In the middle of it all I was invited to join a field trip to Keith's Farm, two hours from the city, as an example of a working sustainable system. I stood among my classmates on a well-traveled, soft path alongside rows upon rows of April garlic shoots, feeling happy to breathe fresh air and listen to the stories of a farmer. I wanted to stay.

My interview centered more around a general evaluation of my ability to persevere under hard conditions than on the details of the job. Luckily, Keith hired me. I had promised Keith and Flavia that I was a giant of sorts, but my lack of athletic ability and love of elevators preyed on me. I began to walk everywhere and insisted on carrying friends' bags home from the grocery store to build the strength that I had guaranteed.

The dreaded test of strength came in the first moments of my first day. The first task: move eighty-pound bags of cement from one place to another. Doubled over, silently praying, I embraced a bag and began shuffling slowly behind the others, who seemed to float with ease to the other pile. Feeling humiliated and doubtful, I kept my two feet shuffling. When my bag began to leak a steady stream, Bill rescued it from me and I took a deep breath.

I never could have predicted the experiences I had there; every emotion that I had ever felt I experienced anew. An acute awareness and sensitivity developed that enabled me to sense minute changes in temperature, to smell cut basil from hundreds of feet away, to hear the scurry of a rabbit, to see a tiny tree frog attached to a zucchini plant, and to taste tomatoes, garlic, onions, watermelon, and carrots directly from our good earth. It drew me closer to the land, to food, to spice, to rest, to joy, to pain, to boredom, to happiness, to frustration, and to love than any experience I can conceive of.

Jessie Burgoyne — Wake up at 5 AM, let Keith's dogs off deer duty, and enjoy a beautiful sunrise. Often I spot turkey, deer, and even coyotes—opportunities not everyone can experience.

Beauty doesn't stop with the obvious—you can also find it in your relations with people who share your passions but possess different personalities. You must learn to take the bad with the good and realize we all come from a different place in this world, geographically and spiritually. Living with five other people has put this belief to the test for me.

It is very hard work, but it's shared, just as we share our love for planting, harvesting, and selling vegetables to people who get just as excited about what we are doing as we do. I've been paid in more than a monetary way to learn. In six months, I've learned more than ever and have had the opportunity to do something that I love. I've loved my work, my new friends (or rather, my new family), and my new surroundings. I've never been so happy in any other place.

It is an effort, and takes perseverance, but we are capable of doing great things in simple ways. This is what this experience has let me do.

Jennifer Fayocavitz — As interns working on Keith's Farm this season, we all have arrived with our own strengths and weaknesses. We have come to work, sweat, dream, eat, sleep, and play for six months. We have brought with us our own histories, personalities, and knowledge. Some of us have worked on farms before; some of us have never seen a wheel hoe (never mind know how to use one).

Though we bring unique contributions to the group, we all do share at least one thing: the intention to be here, to do the work that is required to bring vegetables to the market. United by this, we become part of something greater—a farm that has been producing for many years, a farmer who has had many other interns just like us. In a sense, we become part of a routine, a rhythm that has been ingrained into all that makes up this farm.

My days are long, but the work is fulfilling in a way I have never experienced before. So many worries and concerns have been let go of, and new possibilities have arisen. I am just beginning to see all of this unfold, but I can tell it is going to be quite a journey, worth every tired, aching muscle in my body.

❧ **Kevin Toomey** — Why does the word *farmer* carry a negative connotation in the minds of most Americans? Is it because farmers typically work long hours for very little pay? I guess there is some truth to that: I work forty-five to fifty-five hours per week and I'm definitely not getting rich.

What makes somebody like me want to endure so much for what may appear to be so little? One aspect is the connection between the farmer, the food, and the consumer. Growing organic produce and marketing directly to consumers fills a void that supermarkets cannot provide. The aisles of processed food and the displays of produce with unknown origins leaves one at the lonely end of a complicated food system.

Before I came here, the only organic farming I knew was from exercising my brain. Now I am learning from exercising my body. My fingers have become dry and caked with dirt; for the first month, my right knee had a constant ache that shot down the side of my leg every time I bent over (which I attribute to the bent position that I tend to be in all day). I don't think I have ever paid so much attention to the ground as I have in the last seven weeks. Sometimes I forget there is a sky.

❧ **Stacy Dorris** — My first impression of farmwork was an awareness of how physical it was. My hands began to dry from the dirt and grow callous from using a shovel; my ribs ached from digging; I sweated. The work was hard and needed to be done. I went to bed tired.

Passing by work we had done, I began to notice movement: the tiny lettuces with their new leaves, single green sprouts emerging from the black dirt, blades of grass around the new tractor shed—all work I played a part in. As the days went on, our small group expanded and a rhythm began to establish itself: the work for the day assigned in the morning; new fields prepared and vegetables planted weeks ago revisited to weed, observe, trellis, or water; new vegetables going into the soil in the afternoons.

Working, living, eating, and enjoying all taking place on a farm with five other apprentices and a farmer took a bit of time to adjust to. Like the lettuces and their new leaves, the work and environment is changing me, pushing me. I am growing stronger, more observant, enjoying myself and falling asleep with no problems at all.

Jess Dolan — My room smells of garlic. I smell of garlic. The juice of the plant stung its way into little cuts in my fingers, into my bloodstream, and now it is a systemic odor. It settled into the membranes of my nose so that I both inhale and exhale garlic.

So it is from a dream of beeping garlic that I awake on Saturday-beeping because my sleep-logic imbued the vegetable with a voice like my alarm clock. In the bathroom I aim some cold water approximately at my face. I like to imagine that I am the only one in the world who is awake. But this is not true, and my being awake is only a half-truth: Still surrounded by hypnopompic conversations, I have to keep in mind that it is possible to doze off on the toilet.

At four eighteen I make the short, dark trip from my back door to the truck. Even though it is morning, when everybody, including me, is at their most fragrant—the deep-down odors of slowed organs clashing with the mint and sweet toothpaste and perfumed soaps—the dogs' noses don't recognize me, and their eyes simply can't, and they snarl at me as they would at any wee-hour intruder. But at four twenty, I safely board the truck and am on my way.

Later on, at the market, I talk with a customer who has just bought some garlic. When I reply to her question, what time did I get up, she laughs, "That's when I went to bed." I think of us city and country mice keeping watch for each other, alternately dreaming of garlic.

Ben Lewis — Before I came to Keith's Farm, I lived inside, and was basically oblivious to the weather. But since I moved to the farm, I've begun to listen to the forecast. The weather controls the lives of everyone that works here. It determines whether we'll spend the day planting, weeding, or seeding in the greenhouse. We have to plan when to cut lettuces, pull garlic, and bunch herbs so the plants won't wilt in the heat. Too much sun and we'll have nothing to sell at market. Too much rain and vegetables and morale begin to droop.

I've noticed that my mood tends to follow the cycle of weather through the day. First thing in the morning, it's overcast, the grass is still covered with dew, and I feel sluggish. But soon I'm off to pull garlic, my blood gets pumping, and I start to wake up. As the sun finds its way out of the clouds, we come to the middle of the day. It's humid and the sun is sweltering overhead. Sweat beads fall from my forehead onto the hot dirt as I hunch over to trellis tomatoes. At lunchtime I enjoy a nap on the couch, in front of the fan. In the afternoon, I stay inside to wash mesclun, knowing the sun is still brutal outside. Before too long, evening comes. The shadows grow long and it gets a few degrees cooler. I'm up on the hillside cutting mint. The sweet smell of the herbs wafts through the breeze and a peaceful feeling washes over me. Soon the workday comes to an end. We load the last crates of fresh lettuce and tomatoes onto the market truck as the sun touches the horizon.

Weather affects the farmer every day. It influences everything from deciding when to harvest to changes in mood. In choosing to live and work on the farm, I've become more aware of the sun and rain.

❧ **Stephanie English** — Lately I've been wearing my new identity as a truck driver almost too comfortably. I'll be in my own car, returning from a Port Jervis groceries trip, and as I crest the hill on Route 6, I release the accelerator to let the exhaust brake do its job. It doesn't. I have no exhaust brake on my car. But I will glance down in vain search of its symbol light on the dash before I remember that it is not four thirty in the morning, I am not carting around a ceiling-high truckload of produce, I am not traveling with two co-workers who are dozing to the lullaby of a rumbling Mitsubishi Fuso. Today, I am not a trucker.

At the season's start, we had only enough time for me to desperately absorb three stick shift lessons from Keith before plunging into my Wednesday excursions to Union Square. I began as a Tuesday-night insomniac, terrified by the shiny white box that ensured the futility of a rearview mirror. These days, though, I embrace my market manager responsibility, in all its exhaustion and exhilaration. I love the challenge of marketing absurd quantities of unusually prolific arugula. I feel proud to drowsily return the truck to Keith late Wednesday night, its cargo nearly emptied except for our iced coffee cups. And when customers in hot pursuit of an exquisite salad leave my lovely leafy lettuces upended and rumpled in my display crate, I take it personally.

When my relatives ask if I am still considering grad school, I gladly tell them of my studies: countless workshops in the art of herb-bunching perfection, intensive remedial garlic-sizing school, the full-season seminar in Keith's specialty, "Farming as a Metaphor for Life." We apprentices have many affectionate nicknames for our experience here, but call it what we may, we know well that educational opportunities abound in the life of a farmer. This year I even got a little driver's ed.

✸ **Monica English** – "Monica, could you come here for a minute?" The sun was high over the garlic scapes that we were removing by the crate load, and I was glad for the chance to drop my knife and rest my aching thumb. As I approached, Keith turned toward the garage; I followed. "Are you at all mechanically inclined?" he asked. "Not very," I answered, but his question piqued my curiosity. Seemingly undeterred by my negative response, Keith gestured toward a large, red, intimidating (to me) piece of equipment. This was my introduction to the DR mower, a large walk-behind mowing machine with a single blade sturdy enough to cut through half-inch saplings. Keith went on to explain that proper aeration is crucial in maintaining plant health. To ensure this, he needed someone to take charge of monitoring and mowing the pathways between the crops. Would I be interested in that post? I swallowed and fearfully answered yes.

Why not? He clearly believed in my ability to master the tough-looking machine, and I was eager for responsibility and the opportunity to please my new boss. All the same, that mower seemed to glower at me from the corner of the garage. It was memories of shop-class disasters and a general fear of noisy things that made me doubt my mechanical abilities. But Keith patiently led me through the operation and maintenance procedures. That day I accepted my call to mowing.

Now the DR mower (whom I affectionately refer to as Dottie) and I have a good working relationship. We've had a few difficulties—a missing bolt, a broken oil cap, an unfortunate encounter with an irrigation tube—and I still tend to dislike big, noisy machines, but we get the job done. Taking on this new task has given me satisfaction and confidence, and more importantly, it has made me more aware of all the small jobs and the tremendous orchestration that it takes to run a farm, the many pieces of the whole. That, to me, is what this internship is about. It's about awareness and humility, exploring deeply and understanding systems. I'm learning that farming is full of perpetual challenges and perpetual rewards and that while not all jobs are glorious, on a small organic farm, someone has to mow the weeds.

More than 130 interns have worked on the farm over the past twenty-one years, many for more than one season. Their names are listed in the acknowledgments at the end of the book.

THE UNPEACEABLE KINGDOM

The chickens had lived happily and without serious incident for more than a year. Their routine was well established: down from the perch at first light to take the early sun; a midmorning dust bath; a daily diet of cracked corn, kitchen scraps, leftover salad greens, and the occasional cricket unlucky enough to land within their reach.

No fatalities were recorded, no night raids by foxes or raccoons, no attacks from above by great horned owls. Life in the coop was going along rather nicely, and we humans on the farm were very pleased with the fresh eggs with rich golden yolks left for us each day.

Of course, there had been some infighting (normal in the life of chickens), some small skirmishes among the hens to establish pecking order, and now and then an altercation between the two senior roosters, usually over a hen. The third rooster, recently born to one of the hens on the farm, was young and scrawny, with not much in the way of fighting spurs or tail feathers. He was the odd man out, constantly under attack from both the hens and his fellow roosters. A meek and retiring bird, he seemed to live on the periphery, hoping to snatch a few morsels of food before finding a place to hide.

Compared to most chickens in America, who can barely stand up in their cramped laying cages, our small flock of birds had plenty of room to spread their wings. Their thousand-square-foot enclosure could easily have housed another fifty of them. Even so, the day came when it was not big enough—the grass on the other side was greener. (In fact, the grass on the inside was completely gone due to their incessant scratching and pecking.)

It was Big Red—a large, handsome bird who had fought his way up through the ranks to become the dominant rooster—who discovered that, with a running start, a jump, and a few good flaps of his wings, he could just clear the six-foot-high chicken wire fence that separated him from the larger world.

Within days the other birds followed his example. Big Red was right: It *was* better on the outside. For one thing, there were a lot more crickets; there was also a large compost pile from which to extract small edible creatures. And the hens were delighted to discover many hidden places to lay their eggs.

It was good on the outside, but it was not quite paradise. Also living on the farm were two dogs: Kuri and Bruzzi—one black and the other white, both large, and both given to chasing live, moving objects. And the day came when a young hen ended up in one dog's mouth, badly mauled but still alive. I rescued her, sternly reprimanded both dogs, who were together at the time, and tied them up for two days, hoping to send a clear message.

All seemed to go well for a week or two. But with the summer harvest in full swing and many crops to be planted for the fall, life

It was good on the outside, but it was not quite paradise.

*Attack Pending—
Kuri and the Chickens*

on the farm was busy and there was little time to think about mischief between chickens and dogs.

On a hot, hazy morning in late July Kuri and Bruzzi launched a second attack. This time there was no one around to intervene, and the chickens suffered their first fatality. I was furious and yelled and cursed at the dogs at length, slapped their faces with the dead hen, and tied them up for three days.

It might have ended there. At least two months passed without incident. The chickens continued to leave their enclosure during the day and return to it before nightfall. Then one Sunday morning, a neighbor visited us with her young female dog. Kuri and Bruzzi, both males in their prime who had not yet met the castrator's knife, were intensely interested in any bitch that arrived on the farm, be she in heat or not.

While we were enjoying tea and muffins in the kitchen with our neighbor, the new dog, we surmise, decided to try her hand (or rather her legs) at chicken chasing. This surely would have been too much for Kuri and Bruzzi—their innate lust for blood, their love of a chase, and their desire to impress a young female must have overwhelmed their canine brains and prevailed over any sense of restraint and civilized behavior that I had tried to instill in them. By the time I got outside, the carnage was well under way. Dead and dying birds were strewn about the yard, feathers everywhere, some still floating in the breeze. The injuries inflicted were monstrous, the cries of the wounded chilling. Kuri and Bruzzi immediately retreated to the far reaches of the farm, knowing full well they had transgressed greatly. The neighbor, with profuse apologies, left, taking her little bitch with her.

I counted six dead birds and three or four injured. During the night, two of the latter died. Among the fallen were Big Red and the other large rooster (who had never received a name). The third rooster was nowhere to be found and was presumed dead. That night when the dogs came back, I tied them up without uttering a word and did not feed them. The next morning, with some reluctance, I took the two dead roosters and tied them with baling twine to the dogs' collars. I swatted the dogs a few times with their dead victims, the blows measured to cause shame and humiliation rather than pain. This punishment was repeated every morning for five days, during which time the dogs remained tied up and on light rations. By the fifth day, the two roosters, which were still attached to the dogs' collars, were in a state of decay and smelled accordingly. It was an altogether unpleasant affair.

A week after bloody Sunday, to everyone's amazement, the third rooster emerged out of the high grass along the edge of the driveway. He was limping and was missing a lot of feathers, but, unlike the other two, he was still alive. We named him Lazarus.

Lazarus recovered slowly. He settled in with the remaining hens, his former tormentors, but kept his distance from them. The hens, on the other hand, began to show a friendly interest in him. Whenever possible, they would cluster around Lazarus in a show of support. Gradually he gained confidence. His tail feathers grew, and so did he. He began crowing at daybreak—at first a little unsurely, but soon with full-throated zest. And eventually he was able to furnish the hens with the services they desired.

That was nearly five years ago. Chickens have come and gone, but Lazarus is still here. He has fathered many young chicks and is undoubtedly the patriarch of the coop. All the hens look up to him. Bruzzi, the great white beast, died last year from causes unrelated to chickens. His loss was deeply felt. Kuri is getting older but is still able to run down a woodchuck. Always at my side, he is a loyal and much-loved friend. And to this day, whenever a chicken crosses his path, he diverts his gaze.

Dead and dying birds were strewn about the yard, feathers everywhere, some still floating in the breeze.

THURSDAY
at the FARM

6:00 AM ❧   The alarm goes off. I get up and make my way a little stiffly to the bathroom. Check the weather forecast. Eat a bowl of warmed-up brown rice with a few nuts, raisins, and a banana. Stretch for twenty minutes. Prepare a pick list for the Friday market, looking back at sales figures from the last couple of Fridays.

Turn off the electric fence. Unchain the dogs from their posts in the field. (Both fence and dogs are meant to deter hungry deer during the night, but sometimes don't.)

KEITH'S FARM PICK LIST
HARVEST DATE: 9|17|98

WEATHER FORECAST: okay
MARKET DAY & DATE: FRI 9|18|98

ITEM	CRATES/LUGS
Beans	------
Peas-Sugar Snap	--
Peas-Snow	------
Broccoli	------
Broccoli Leaves	--
Collards	------ 1.5
Kale-Winterbor	--- 1.5
Kale-Red Russian	- 1.5
Kale-Lacinato	-----
Kale-Redbor	-----
Mustard-Green	------ 1
Mustard-Red	----- 1.5
Turnips-White	----- .5
Turnips-Gold	----
Turnips-Purple	--- .75
Tatsoi	------- 1.5
Mizuna	------- 1.5
Sorrel	------- .5
Dandelion C/P	---- 1.5
Endive	------
Escarole	------
Lettuce-Fr Crisp	- 4.5
Lettuce-Red Oak	-- 4.5
Lettuce-Romaine	-- 2.5
Lettuce-Other	------
Garlic	------- 1.5 large
Onions-Cooking	-- 1
Onions-Salad	--- 1
Scallions	----- .5
Shallots	------- .5
Celeriac	-------
Peppers-Sweet	---- .5
Peppers-Hot	--- .25
Potatoes-White	-- 1
Potatoes-Yellow	-- .5
Potatoes-Red	---- .75
Potatoes-Blue	----
Carrots	------- .75
Radishes	------ .75

ITEM	CRATES/LUGS
Tomatoes:	
Cherry-Red/Orange	1/2
Heirloom	------ 2
Red-Large/Small	- 5
Plum	------- 1
Green	------- .5
Mesclun-Coolers	-- 3
Arugula-Coolers	-- 1
Swiss Chard	----- 1.5
Ppt Spinach	-----
Summer Squash	---- 1.5
Zucchini	------- .5
Winter Squash:	
Acorn	----------- 1
Butternut	--------- 1
Delicata	---------
Sweet Dumpling	-- 1
Other	-----------
Pumpkins-Lge/Sm	-- .5
Pumpkins-Mini	----

Herbs:	BUNCHES
Anise Hyssop	---- 5
Basil	----------- 35
Catnip	--------- 5
Chives	---------
Cilantro	---------
Lavender	--------- 10
Lovage	--------- 10
Marjoram	--------- 10
Oregano	--------- 40
Parsley-Italian	- 20
Parsley-Curly	---
Rosemary	--------- 15
Sage	----------- 10
Savory S/W	------ 4
Tarragon	---------
Thyme-Reg/Lem	--- 12/6
Mint: Spear.	------ 6
Choc.	-----
KC	-----
Julep	-----
Orange	----
OTHER:	

OTHER WORK

• Move sprinklers to 06 brassicas

• Drip irrigate basil

• Rototill for last lettuce

Pick List for Friday Market at Union Square, 9/18/98

8:00 AM ❧ Meet with the crew (Matt, Laura, Graham, Ione, Jessie, Sandra). Exchange a few words of greeting, then a quick preview of what we need to accomplish in the day ahead. Each crew member receives his or her pick list for the morning. Friday is the smallest of our three markets, so the pick is relatively light. We hose out the number of lugs and crates that we think we'll need, load them onto the old Dodge pickup, which is no longer registered for use on the open road, but serves well as a farm harvesting truck, sharpen our knives, and head for the field, each to his or her assigned tasks.

For an hour or two we work quietly, filling lugs with kale, collards, Swiss chard, lettuces, mizuna, tatsoi, carrots, squash, zucchini, dandelion, mustard greens, potatoes, onions—whatever is ready for harvest and looking good.

10:15 AM ❧ Jessie hears a faint bleating sound in the pasture behind Field 05. Matt and I climb over the barbed wire fence and after a few minutes stumble upon a baby calf hiding in a thicket. It is lying down, not much more than skin and bones, and certainly too weak to walk. We carry it back to the barn and give it water. It drinks and drinks and drinks while we stand and watch, touched by its gentle, frightened eyes and tenuous hold on life.

I call Ronnie Odell, the dairy farmer who rents our pasture, and he comes right over to claim the little foundling. Its mother is back

Newborn Calf

at his place. He had taken her home in a trailer a week earlier, expecting that she would have her calf in the safety of his own barn, only to find out that she had given birth prematurely. Ronnie had spent several hours searching our pasture but had not found the calf and had given it up for dead. Now here it is. After seven days alone and unnourished in the world, its newfound life is a gift.

Back to the vegetables. We spray everything we've just picked with cold water and then load the lugs into an air-conditioned cooler, where they will sit for the rest of the day. As we work, we munch on a few freshly dug carrots and maybe a Japanese turnip or two. Graham peels a clove of garlic and chomps it down raw, with a devilish grin on his face.

11:00 AM ❧   Time to get the mesclun. We head up the driveway with a dozen plastic tubs and set about cutting an assortment of baby greens in Field 08. A couple of turkey vultures are circling high overhead—there must be something dead in the vicinity. The sun is getting hotter, and some of us are beginning to feel a little late-morning fatigue. But we keep going, and an hour later the mesclun is all cut and each of us is ready for a break from bending.

We're glad for the mesclun washing. We do it in big, white laundry tubs in the old milk room, which is pleasantly cool. The radio is on, or maybe a Bob Dylan tape. We chat, relax a little, engage in some light banter.

1:00 PM ❧ Lunchtime. And we're ready for it. I eat a sandwich, drink a cup of sage tea with fresh ginger in it, and jot down a few "must-dos" for the afternoon. Then, something I've been looking forward to for the last hour or so: a few minutes of reading followed by a nap—a sweet, undisturbed, twenty-minute-long nap.

2:15 PM ❧ I get a call from the shop steward at Angelica Kitchen—perhaps New York City's finest organic vegan restaurant. I tell him what vegetables and herbs we have and which are at their peak of quality and flavor. He considers his menu for the next couple of days and places an order that includes plenty of greens, some Sungold cherry tomatoes, carrots, parsley, garlic, rosemary, thyme.

After lunch, two or three of us continue washing mesclun while the rest set about filling the Angelica Kitchen order. Then we move the irrigation sprinklers and connect up drip lines in the basil. I get in a half hour of tractor work in preparation for some lettuce planting. The afternoon moves fast. There's lots to do and not quite enough time in which to do it.

4:00 PM ❧ We still need to get tomatoes and peppers. We need to cut parsley and other herbs and we need to mix the mesclun, which by now is washed and dried and free of stray leaves of pigweed or offending blades of grass. And the Angelica Kitchen order needs to be weighed and carefully packed and labeled.

7:00 PM ❧ Sandra starts filling out the Harvest and Sales Record Form for the day, noting how many crates, coolers, or bunches of each vegetable and herb are going to market and what field they came from (this latter entry is required by our organic certifier). She quizzes her co-workers to be sure she's getting the correct

KEITH'S FARM HARVEST & SALES RECORD

HARVEST DATE: 9/17/98   MARKET DAY & DATE: FRI 9/18/98   MKT WEATHER: *Good*

ITEM	FIELD	AMOUNT	AMT SOLD
Beans			
Peas-S.Snap			
Peas-Snow			
Broccoli			
Brocc Leaves			
Collards	05	1.5	AU
Kale-Wintbor	06	.5	1.25
Kale-Red Russ	06	1.5	AU
Kale-Lacinato			
Kale-Redbor	06	1	.75
Mustard-Green	06	1	.5
Mustard-Red	03	.5	AU
Turnips-White			
Turnips-Gold	03	.75	.5
Turnips-Purpl			
Tatsoi	05	1.5	AU
Mizuna	05	1.5	1.25
Sorrel	08	.5	AU
Dandelion C/P	05	1	AU
Endive			
Escarole			
Lettuce-Fr Cr	01.1	4.5	4
Lettuce-R Oak	01.1	3.5	AU
Lettuce-Rom	01.1	2.5	AU
Lettuce-Other			
Garlic	01.3	1.5L	1
Onions-Cook	01.2	1	.75
Onions-Salad	01.2	1	AU
Scallions	01.2	.5	AU
Shallots	06	.5	.35
Celeriac			
Peppers-Sweet	07.2	.5	.4
Peppers-Hot	07.2	.25	.2
Potatoes-White	06	1	.75
Potatoes-Yelo	06	.5	.3
Potatoes-Red	06	.75	.6
Potatoes-Blue			
Carrots	06	.75	.5
Radishes	01.2	.75	AU

ITEM	FIELD	AMT	AMT SOLD
Tomatoes:			
Cherry-Rd/Or	01.2	1/2	AU
Heirloom	01.2	2	AU
Red-Lg/Sm	01.2	5	4
Plum	01.2	1	.5
Green	01.3	.5	.25
Mesclun-Coolers	08	3	2.5
Arugula-Coolers	08	1	AU
Swiss Chard	01.1	1.5	1.25
Perp Spin			
Summer Sq	01.2	1	.9
Zucchini	01.2	.5	AU
Winter Sq			
Acorn	09	1	.5
Butternut	09	1	.75
Delicata			
Sweet Dumpling	09	1	.75
Other			
Pumpkins-Lg/Sm	09	.5	.3
Pumpkins-Mini			
Herbs-Bunches			
Anise Hyssop	08	5	4
Basil	01.3	35	AU
Catnip	08	5	3
Chives	08	8	AU
Cilantro	08	8	AU
Lavender	08	8	6
Lovage	08	10	AU
Marjoram	08	8	5
Oregano	02	10	7
Parsley-Ital	05	40	35
Parsley-Curly	05	20	15
Rosemary	02	15	14
Sage	02	10	9
Savory S/W	08	4	AU
Tarragon	02	6	AU
Thyme-Reg/Lem	02	12/6	AU/4
Mint:Spear	02	6	5
Choc	08	8	AU
KC	08	8	AU
Julep	08	6	4
Orange			
OTHER:			

Harvest & Sales Record for Friday Market, 9/18/98

numbers. They may differ in small measure from the numbers on the Pick List, since the pickers have some leeway to harvest a little more or less, based on availability in the field. At the end of the market day, the amount of each item sold will be entered into a column next to the "Amount Taken or Harvested" column. These harvest and sales figures, along with weather data, are invaluable when it comes to deciding how much to pick for subsequent markets.

The sun is low in the sky and the temperature is just right. We're ready to hose out the milk room and start packing the big box truck that will go to market. A dozen barn swallows, most of them born on the farm this season, are sitting together on the electric line that feeds the barn, and the pigeons that roost in the silo are coming in for the night—they descend in wide, circular patterns, almost as though they're floating through the still air and taking pleasure in their last aerial experience of the day. We laugh and joke as we load the truck, feeling good about the work we have done, looking forward to the rest that the night will bring. And maybe there's still time for a quick dip in the pond.

◆

Friday, 4:15 AM ᔍ   Laura and Ione will head out for the farmers' market in Manhattan. The forecast is good, so they should have a decent day. The rest of us will stay back at the farm and harvest for Saturday, a bigger market. It's a busy life.

WILD WEATHER

(August 2000)

Everybody's talking about it—the wild weather we've been
having: heat and drought in the south, heat and fires in the
west, cool temperatures and rain, rain, rain in the east. We've had
more rainy days and, in some areas, more rain, period, than any of the
old-timers can recall. For farmers in the Hudson Valley and upstate
New York, the old adage "Make hay while the sun shines" was never
more pertinent than in the summer of 2000. (It's essential to cut and
bale hay while it is relatively dry and ideally still green. Wet hay gets
moldy and is not good for cows and horses. It also can spontaneously

combust and burn down your barn. If you slept in on a dry day when you might have been cutting hay, you probably lived to regret it. Hence, "Make hay while the sun shines.")

Farmers of all stripes, not just hay cutters, have been affected by the cool, wet summer. One vegetable grower I know, John Gorzynski, who farms in Cochecton Center in Sullivan County, New York, got ten inches of rain on a Saturday afternoon in August and three inches the day before. On that one Saturday Gorzynski lost about 80 percent of his remaining production for the year. One of his greenhouses collapsed, and the flooding was so severe in a low-lying field that eight inches of topsoil got washed away by moving water. That kind of weather sets you back.

Butternut Squash

Last summer was dry and hot, and farmers without irrigation equipment and plenty of water did not fare well. This year irrigation systems lay idle and many farmers watched crops rot in the sodden ground. In June and July 1999 we got 0.85 inches of rain on our farm. In the same two months of 2000 we got over twelve inches.

There was so much water in one of our fields that a pair of Canada geese and their goslings thought for a while they had discovered a new pond. I set up a pump and drained the water, disappointing the geese but pleasing the onions that were growing there. Most of them survived and grew to maturity, but now, after harvest, we're seeing a high incidence of neck rot. Same thing with our shallots.

The winter squash and pumpkins got off to a late start and, due to the cool weather, are not flowering and fruiting as they should, which means we won't have many squash to sell this fall. And some of our Mediterranean herbs, most notably rosemary, sage, and thyme, are looking pretty sad from having wet feet for so long.

But not everything has gone badly. The rain in May and June made our garlic grow nicely, giving us large bulbs, though the rain in July created a harvesting crisis. If you want garlic that will store well and taste good for six to eight months (and we do), you need to cure it by letting the tops dry down. Garlic harvested in wet conditions is difficult to cure. But leaving the bulbs in the ground too long, waiting for dry weather, means they will keep growing and soon burst out of their skins. You'll be left with big bulbs with exposed cloves that tend to break apart and have a short storage life.

We gave up on waiting for dry weather and brought the garlic in wet, something I've never done before. Within a couple of days, white spores started appearing on the damp and still partially green tops. With much trepidation, we cut the tops off and stood the bulbs one layer deep in ventilated crates. We then turned on the fans and have been blowing air across them day and night for more than a month, hoping to avoid bacterial invasion. It seems to be working—but it's using a lot of electricity and a lot of fans. In one anxious week I bought seven new fans, including one big, belt-driven thirty-six-incher for $475.

Farmers tend to grouch a lot about the weather. It's one of those things that seldom works out just right for us. And if it ever did, you can be sure we wouldn't let on. This year there was more than just rain to contend with. Apple growers in the Hudson Valley got struck by one ferocious hailstorm after another. A farmer in Ulster County spoke of wind-driven hailstones penetrating his immature apples like bullets from a gun. Many orchardists are facing huge financial losses; it is estimated that two million bushels—at least a third of the apple crop—is not worth harvesting.

The hail came at a bad time. Hudson Valley apple growers are already in a tight spot, facing depressed apple prices and small profit margins due to a glut of apples on the market worldwide. They are also under intense development pressure. No doubt we'll be trading in more orchards for subdivisions in the near future.

Being diversified can be a big help. On our farm we've had better lettuces this year because lettuces and other greens do well when it's cool and damp. Tomatoes, on the other hand, taste best when they ripen on hot, sunny days—cool, overcast weather and too much water robs them of their sweetness and promotes the spread of fungal diseases.

While not averse to flattery, I always feel a little uneasy when an enthusiastic customer comes to our stand and says something like: "Keith, those orange cherry tomatoes I got from you two weeks ago were the absolute best I've ever eaten. Better than candy." I fear the compliment is like a debt that I might have to repay. An intervening cool, wet spell can take the flavor out of cherry tomatoes unbelievably fast. Too much rain also makes them prone to split in the shopping bag on the way home. A happy customer can turn into one who is quite unimpressed in just a matter of weeks.

Many people think it is our God-given right to have a full range of fruits and vegetables available to us at all times. Seldom do they give much thought to where the food comes from, what it took to grow it, or the fact that the farmer in the field often is not in full control. True, we can strive to build healthy soil, maintain fertility, dodge insect pests and plant diseases, and sow and harvest crops at the optimum times. But beyond these areas of human influence, there are the sun, the wind, and the rain. These three elemental forces (and variations on them, such as hail, blizzards, and early frosts) can have profound effects on the size, appearance, and flavor of what we harvest and what you ultimately eat.

It's different in some parts of the country. In California, Arizona, and Florida, sunshine, irrigation, and chemicals produce more uniform and consistent crops—crops that may start out looking good but that lose much of their luster (and nutritional value) during the cross-country travels and multiple handlings they must endure before arriving on our plates.

In the Northeast, the weather rules. We farmers are more like hunters and gatherers, not knowing what the next day or the next week will bring. But there's something to be said for not always being in full control, for not having nature right under your thumb. The wild and erratic weather that we so like to gripe about and that sometimes strikes us without mercy is often enlivening. It keeps us on our toes and reminds us that we're not the only force to be reckoned with on the planet. And it is a thread of communality that binds those of us who work on the land. Speaking only for myself: except when I'm especially disgruntled after a long, wet week, I wouldn't have it any other way.

Farmers tend to grouch a lot about the weather. It's one of those things that seldom works out just right for us.

Cipollini Onion

SMALL-FARM ECONOMICS— watching the bottom line

What's it take to be a small farmer these days? Good health and a sound body. A practical mind. A farm in the family or enough capital to buy one and get started. A passion for working with the land. The ability to market your own products. The will and stamina to put in seventy or eighty hours a week when necessary. A gambler's instinct. A willingness to accept the slings and arrows of outrageous weather.

All these things are important and will help to give you a fighting chance. But what today's farmer must have, above all, is a good nose for business and the bottom line. To be successful you have to make enough money to pay the bills and still have some left over. Otherwise creditors will be at your door, your spirit will wither, and you won't last!

Many farms fail because too much energy goes into farming and not enough into analyzing the inflow and outflow of cash. It's nice to think of the farmer as a natural man (or woman), close to the land, untainted by the ways of the commercial world. But farmers like this live more in the public's nostalgic imagination than in the real world.

With vegetables you make your money one dollar at a time. A head of lettuce for $1.75, a pound of carrots for $2.00, a bunch of parsley for $1.50. You must learn to handle volume and handle it efficiently. If you can find a way to cut and rubber band a hundred bunches of parsley in an hour instead of eighty-five, then you're heading in the right direction. Those extra fifteen bunches are where your margin of profit lies. Having a beautiful bunch of parsley is important. But it's not enough. You need a hundred of them.

The annual cost of running a farm tends to go up from one year to the next. It seldom goes down. But the farm's productivity and income are much less predictable. A few years ago many of the onion growers in the black-dirt region of our county were hit by a freak hailstorm. The damage to the young plants was severe, and some of the farmers affected were forced out of business. That spring also brought unusually heavy and prolonged rain. Eighty percent of our potatoes rotted in the cool wet soil before they had a chance to sprout. The cost of the potato seed, the ground preparation, and the planting were about the same as last year. But when we sold the potatoes, the return was small—it barely covered the cost.

However, unlike the vast, black-dirt onion fields to our east, our farm is highly diversified—we grow more than one hundred different varieties of vegetables and herbs in a single year and tend to some twenty-five fruit trees. We sell directly to the public, never wholesale. No middlemen siphon away our profit.

The early rain that was so damaging to our potatoes was just right for our garlic, which was planted on a well-drained slope.

We sell directly to the public, never wholesale. No middlemen siphon away our profit.

Garlic likes plenty of moisture early on and then relative dryness later, which is exactly what it got. The bulbs were big, they tasted good, and they stored well. That year we made money on our garlic.

Which brings me back to my claim that a gambler's instinct is not a bad thing for a farmer to have. At the beginning of the season the farmer places his bets. How much will he or she plant of each crop? Will there be multiple and/or consecutive plantings? Where will each planting go, and when? There are many variables to consider: labor costs, weather, insects, disease, marauding deer, oversupply in the marketplace, and fluctuations in demand. If you bet on just one item and something goes wrong, you may be in big trouble; if your bets are spread over a wide variety of vegetables and herbs, so is your risk. Diversification gives a farmer more options. If one crop shows signs of failure, you can choose to let it go and direct your energies elsewhere.

So what are the bottom-line costs of running a farm and what is the potential for income? Doubtless a hundred different farmers would give you a hundred different answers—no two farms are exactly alike. I can only speak about our farm.

We grow a wide variety of vegetables and herbs organically on about twelve acres. Twice a week, during the growing season (we recently switched from three to two days a week), we truck our produce to the Greenmarket at Union Square in Manhattan. On a good day we will sell most of it. On a bad day we may sell only two-thirds to three-quarters of what we bring. We can make eight hundred dollars at a market or four or five thousand, and sometimes more—depending on the day of the week (Saturday is better than Wednesday), the time of year (fall is better than summer), the weather (rain, cold, and oppressive heat keep customers away), and what we have to offer (there's much more to sell in the fall than in spring or early summer). The Greenmarket has strict "grow your own" rules which we adhere to and support. This means we never supplement meager farm production with purchased vegetables from a wholesale market.

A gambler's instinct is not a bad thing for a farmer to have.

Our expenses are many. Labor is the biggest. In order to farm twelve acres effectively and because organic farming is highly labor-intensive, we need six or seven seasonal workers. Labor and associated insurance and payroll taxes are usually around $60,000. Here's a breakdown of major non-capital-improvement expenses for an average year, post-2006 (figures are rounded off):

Labor, workers' comp, etc.	$95,000
Insurance (vehicles, farm)	$6,000
Seeds and plants	$3,000
Greenmarket fees	$7,500
Gas, diesel, oil	$5,000
Tolls	$1,000
Potting mix, lime	$1,200
Organic fertilizer, manure	$1,500
Organic certification	$1,500
Crop expenses	$2,500
Property taxes	$6,500
Supplies	$2,500
Equipment maintenance & repair	$3,000
Building maintenance & repair	$2,000
Accountant fees	$2,500
Office, telephone	$1,500
Utilities	$2,500
Postage, shipping	$300
Miscellaneous	$4,000
TOTAL	**$149,000**

In addition, every two or three years we have a large capital-improvement expense, such as buying a $10,000 tractor or a $20,000 truck, or digging a $9,000 pond for irrigation, or building a greenhouse or a shed to house tractor implements. Several years back we upgraded our farmworker housing at a cost of about $50,000. That was a big one, but we felt it was necessary to make our farm an attractive place to work and to ensure its long-term viability. Much of the money we spend goes into the local economy and we feel this is a good thing.

Small farms may not always produce food as cheaply as the agribusiness giants do, but so much of the cheap food we are accustomed to in America comes at a high price when you factor in the social and environmental costs: depressed rural communities, pesticide-related health-care costs, impoverished soils, depleted groundwater, and an end product that is short on freshness, nutrition, and vitality.

Small farmers provide a product with more integrity. We are a link between the buying public and the land. And often we farm for the love of it. But to come out ahead in the competitive food market, we need to pay attention to the bottom line, we need to plan carefully, we need to listen to our customers and treat them well, we need to constantly look for ways to become more efficient, and, in the final analysis, we need to sell a lot of bunches of parsley.

A GARLIC AFFAIR

Of the approximately one hundred varieties of vegetables and herbs we grow on our farm, garlic reigns as the sovereign queen. I would give up the ninety-nine others, albeit reluctantly, before I would give up my garlic. Garlic is our biggest crop and the one that has brought us major press coverage, both in New York City and nationally. Finally and, perhaps most endearingly, garlic is the crop that brings in the most cash.

Most growers of garlic, be they weekend dabblers or for-profit players like myself, soon learn that they have entered into a relationship with a plant that will not be easily cast off. Garlic's attributes are such that, once smitten by the garlic bug, many growers develop a lifelong attachment. Often, our passion for *Allium sativum* goes well beyond its wondrous culinary, medicinal, and curative properties. For me, it is the plant itself that is most remarkable: its stately appearance in the field, its fascinating life cycle and growth habit, its hardiness, its ancient lineage, the way it comports itself in this world.

Garlic is believed to have originated in the foothills of mountainous south-central Asia (northern Iran, Afghanistan, northeastern Turkey, China). It probably was one of the first wild plants to be cultivated by humans, going back perhaps ten thousand years. We can imagine precious bulbs of garlic being carried along silk trade routes by nomads and hunter-gatherers long before silk was being traded. Today, garlic has found its way to all corners of the globe and its many cultigens have adapted brilliantly to diverse climates and soils.

Traditionally, the center of garlic production in the United States has been Gilroy, California. The vast bulk of garlic grown for processing and supermarket sales comes from this area. Gilroy's garlic is predominantly of the softneck type. Softneck garlic is relatively easy to grow on a large scale and stores well. But the bulbs of softneck garlic have numerous small cloves that overlap one another and are often irksome to peel, and the flavor, while adequate, is rarely exceptional.

The Northeast, with its cold winters, is better suited to growing hardneck garlic, a different subspecies that is closer to the original wild garlic from south-central Asia and not as domesticated as the softneck varieties. Hardneck garlic (sometimes called topset garlic) has larger cloves that radiate out from a hard central stem. They peel easily and their flavor, while it ranges widely from one hardneck variety to another, is often outstanding. Hardneck garlic is more demanding to grow, tends to yield less per acre, and often has a shorter shelf life, but among real garlic lovers it is the only stuff to eat.

Rocambole Garlic

On our farm we grow Rocambole, a variety of hardneck garlic that arrived in my hands seventeen years ago through good fortune and the generosity of a neighbor. Andy Burigo, an old Italian American who lives down the road from us, befriended my wife, Flavia (also of Italian ancestry), while she was out on one of her landscape-painting excursions. After Mr. Burigo learned that I was running an organic farm, he presented my wife with a brown paper bag containing about thirty bulbs of garlic and suggested to her that I try growing them. He told her the original planting stock came from Calabria, Italy, and that it had entered the United States many years earlier in the pocket of a friend, unbeknownst to customs officials. Ever since then he had given it pride of place in his extensive and well-tended garden.

That fall I separated the bulbs Andy Burigo had given me into a couple of hundred cloves and planted them in fertile soil. They lay in the ground all winter with a blanket of straw mulch covering them. At the end of March they emerged as the first crop of the season—slender, blue-green shoots that grew quickly. Over the next few months I provided water and pulled the weeds that competed with them. At the end of July, after their leaves had started turning brown, I dug up a couple of plants, brushed the soil off them, and beheld a marvelous sight. The bulbs had a vibrant aura about them. Their buff-colored skins were streaked with a reddish-purple blush. They were firm and well formed. And they were big.

I would give up the ninety-nine others . . . before I would give up my garlic.

Though this was not the first garlic I had ever planted and harvested, that day marked the beginning of my perennial romance with the "stinking rose." I gave the bulbs to my wife. She used them in a meal that night and told me it was the best garlic she had ever eaten. Though not generally regarded by others, or myself, as a man with a discerning palate, I was inclined to agree.

A couple of days later I sold a few dozen bulbs at my farmers' market stand. The following week almost every customer who had purchased one came back smiling, asking for more. It occurred to me then that I might be on to a good thing. I didn't sell any more garlic that year and was reluctant even to give the occasional bulb to my wife. Instead, I squirreled away the hundred-odd bulbs that were left. That fall I divided them into about eight hundred cloves and planted them with great care.

The following year we sold a few hundred at market, again to rave reviews, and saved the rest for planting. I continued like that for a while, parceling out my trove in a quite parsimonious fashion, but within a few years I had built up a planting stock of twenty thousand cloves and an eager pool of customers. I was ready to do serious business.

With each subsequent year, aided by good press (on TV, radio, and in print media), the demand for our garlic has increased. And each year, to keep pace with this demand, I have allotted more acreage and labor to the cultivation of this exceptional plant.

Now we are planting approximately sixty thousand cloves—each and every one by hand. It may be that we have reached a natural ceiling in garlic production, if not in terms of how much we can sell at market, then in terms of the resources we have available to grow the stuff. These days I often feel overwhelmed by the vast sea of garlic growing in my fields and the substantial effort required to plant it, mulch it, weed it, water it, harvest it, cure it, grade it, and sell it. But I still dearly love my garlic and regard it more than ever as the plant that defines the essence of our farm.

Growing sixty thousand garlic plants on a small, diversified, organic farm is no small task. It must be approached in a highly organized, almost military fashion. At each stage, timing is critical. First we select the planting stock—some eight or nine thousand of our best bulbs from the summer harvest. We prefer large bulbs, but not the very biggest—these have a high proportion of split cloves that grow two or three small plants instead of one large one.

In early October we look for a warm and comfortable spot and sit down to separate the chosen bulbs into their constituent cloves. This phase of garlic planting is known as "clove popping." We grade the cloves into several categories (tiny, small, medium, large, questionable, and "bad stuff") depending on their size and quality. The few cloves that are soft, moldy, damaged, or exhibit even the slightest sign of disease go straight into the "bad stuff" box and are later burned in a 55-gallon drum. I regard it as imperative that our planting stock be clean and well screened. The whole process usually takes two weeks and is quite taxing on the hands, especially the thumbs. By the time all the cloves are "popped" and ready to plant, my helpers are wondering if our workers' compensation insurance covers thumb-replacement surgery. (It does not.)

Next, I use a tractor to cut furrows eighteen inches apart in well-rested ground. We then set about on our hands and knees, planting the cloves one at a time at a spacing of three to six inches. Each clove is pushed a few inches into the soil and must be oriented correctly, so that its first shoot in the spring will head toward the sun, not the earth's molten core. The smallest cloves are planted more closely (they receive the three-inch spacing) and will be dug and sold as green garlic in early summer—the entire plant is sold, leaves and all, to the surprise and bemusement of my newer customers. The larger cloves receive five or six inches of spacing and will be allowed to grow to maturity.

After planting is completed (it usually takes two to three weeks), the cloves are mulched a few inches thick with a hundred-odd tons of well-aged bedding material from a nearby horse farm. Through the winter, they rest in the cold ground, nursing their store of energy, awaiting the transformation to come. For a farmer it is a good thing to know that the garlic is in the ground, that the next generation of this most special plant is waiting under the snow to be born.

The first green shoots break ground in late March or early April, and that's when I know for sure that I'm back in the garlic business. By mid-April all the plants should be up. May and June are months of intensive weeding, much of which is done by hand. If the rains fail, water will have to be provided via irrigation.

The garlic carries on its aboveground growth rapidly until the summer solstice, when the longest day of the year is reached. As the days begin to shorten, the plant slows down its photosynthetic processes and begins to focus on its underground parts—the energy captured in the leaves is directed downward to form the new bulb.

Toward the end of June our garlic sends up a flower stalk, though it's more correctly referred to as a false flower stalk, since garlic rarely, if ever, reproduces sexually (via the coming together of male and female parts), like most other plants do. Instead, its strategy for self-perpetuation relies on clonal division: Each new bulb is a clone of an earlier bulb, going all the way back, you might even say, to an ancient mother bulb from some distant time and place.

The false flower stalk of our Rocambole garlic, if left on the plant, can grow two or three feet high. It goes through some wonderful loops and whorls and eventually straightens up and swells at the top to form a capsule that contains several miniature balls of garlic known as bulbils. Most growers believe that early removal of the false flower stalk—the scape or top, as it is often called—will lead to a larger bulb. We subscribe to this belief, too, but we usually leave some tops on anyway since they make such a sight in the field, and, later, can present a stunning arrangement in a vase

The growth of the false flower stalk, the development of the capsule, and the formation of the bulbils are all part of what makes hardneck garlic such an extraordinary plant. Visually, the tops are striking. They are also excellent to eat.

We harvest our garlic when about half of the leaves have turned brown, usually over a two-week period from late July to early August. This calls for major effort on the part of all hands present and generates a copious amount of human sweat. A tractor with cultivator tines loosens the soil on either side of the bulbs so that most of them can be pulled by hand, without additional digging. But the numbers are great, the sun is hot, and the total harvested plant weight, along with a little residual soil around the roots, is several tons.

With their leafy tops still attached, the bulbs are hung in clusters of ten or twelve in every available space in the barn and tractor or implement sheds. Strategically placed fans assist in the curing process. Access into these large enclaves of hanging plant matter is severely limited, and everywhere the air is redolent with the smell of fresh garlic. If the weather is not too humid, within a month the stems of the plants will be sufficiently dry and hard that the leafy tops can be cut off without risk of bacteria entering the bulbs. Once the leaves are removed (this is typically carried out

Dormant in the frozen ground, the next generation of garlic is waiting to fulfill its ancient destiny.

over several weeks), the bulbs are graded according to size and quality. The largest bulbs are usually sold first. They are prized by our customers but do not store quite as well as the smaller and medium-size bulbs.

If all goes according to plan, by December we are taking our last bulbs to market (with the exception, of course, of a personal stash) and our customers are stocking up for winter. Meanwhile, dormant in the frozen ground, the next generation of garlic is waiting to fulfill its ancient destiny and, at the same time, keep its promise to help a small farm stay afloat. It's not a bad deal, on both sides.

◆

A FEW YEARS back Andy Burigo, gardener extraordinaire and father of our garlic, died at the age of eighty-six. In his later years he and his wife, Ida, visited our farm once a year or so to have coffee with Flavia and evaluate the condition of my crops. On their last visit, by which time I had transformed his original gift into some thirty-five thousand healthy plants glistening in the morning sun, Andy called my wife over to him, saying that he had something for her. With a sober look in his eyes, he reached out and pressed into her hand a half-dozen newly minted pennies. "See what your husband can do with these," he said, and then broke into his customary twinkle-eyed laughter.

At his well-attended funeral on a Saturday morning, while I was selling garlic in New York City, Flavia reached into her handbag and took out a very large bulb I had given her the night before. Quietly she made her way through the crowd of mourners and placed the bulb on Andy's coffin just before it was lowered into the ground.

BARN
SWALLOWS

E very year, toward the end of April, the barn swallows return.
It always gladdens my heart to see them. They are such perfect,
tiny travelers from so far away. I have learned to anticipate their arrival
by a week or two and remove a few of the sash windows from the lower
barn so that they will have ready access to their seasonal lodgings.

The first pair to arrive lays claim to the most attractive real estate
we have to offer—one of several mud nests attached to the barn rafters
just below the ceiling, well out of reach of prowling cats and curious
humans, and preferably near an open window. Soon a second and

third pair arrive and take up residence, and the barn is all aflutter with these immaculate creatures with their long, forked tails, their chestnut-colored bellies, and their iridescent blue-black wings.

Swallows build their nests with mud and reinforce them with straw and hair and any other binding material they can find. It takes hundreds of small mud pellets and several days of work to construct a single nest. The nests are invariably attached to vertical surfaces close to the ceiling. Rough lumber seems to be preferred, or else surfaces with some texture that allow the mud to more easily adhere. One winter, I installed new overhead light fixtures in the barn and the next spring these became favorite spots to locate new nests. Fearing electrical shorts and fires, I removed these nests in the fall after the swallows had gone and wrapped the light fixtures with a protective covering of chicken wire. Since then, one or two enterprising birds have found that chicken wire serves their purposes just as well as light fixtures.

One fall, after the swallows had gone, we hired two men to clean and whitewash the interior of our barn. They used a high-powered compressor to blow away dust and cobwebs and chips of paint, and in the process knocked down all the swallow nests. I was concerned that the birds might not come back the next spring. But they did, and within a couple of weeks there were three sturdy new nests attached to the rafters, each one weighing several times the sum total of its residents. Since then more nests have been constructed, offering new arrivals a variety of spots to choose from.

Checking the occupied nests with a mirror in late spring, I will find four or five whitish-brown, speckled eggs in each of them, arranged like little clusters of elliptical marbles. I don't know if the sitting duties are shared by the male and female members of each mated pair, but once the eggs hatch it is clear that both parents expend considerable effort to feed their young. All day long they fly the open sky in search of food to bring back to the hungry hatchlings, who remain in the nest for about three weeks.

Mosquitoes, moths, flies, beetles, and caterpillars are all part of a swallow's diet, and a diversified vegetable farm is an ideal spot in which to find these tasty morsels. Of course, from my point of view, the swallows are welcome guests—what farmer would not be delighted to have this natural airborne pest control? Barn swallows have large appetites and consume great quantities of flying insects, and more than a few crawling ones, every day.

These birds have an unsurpassed talent for aerial maneuverability. They swoop and glide and turn and dive with impressive speed and grace. They love to follow behind the tractor when I am mowing or engaged in any work in the fields that disturbs and flushes up insects. On one occasion I was so taken with their aerobatic displays that I almost drove my tractor into a ditch.

On a still day, at dusk, the swallows catch mosquitoes over the pond, sometimes swooping right down to the water's surface, creating a tiny splash and appearing to take a quick dip and perhaps a drink to go with their evening meal.

Barn Swallow

By midsummer the three or four pairs that arrived in April have multiplied into as many as twenty-five or thirty birds. I have watched the young fledglings lined up on a barn windowsill, eager to test their wings but not quite ready to take on the larger world. Outside the parents flitter back and forth, offering flying lessons and chattering encouragement.

Each pair of adults usually raises two broods of young in a season. At this rate one might imagine the world soon overrun by barn swallows. I can think of worse fates, but, in any event, this

*Swallow Nest and
Hanging Garlic*

does not happen. Many hazards befall these small birds. Some may live seven or eight years, but most perish before the end of their first year, usually from cold or starvation.

But, as a species, barn swallows are great survivors. Wisely, perhaps, they have learned to adapt to a life in close proximity to humans and to offer pest-control services in exchange. They are found throughout North America and most other continents of the world. They spend their summers in temperate zones and migrate to warmer climes in the winter. The swallows on our farm might fly to Mexico or Central America or even as far south as Argentina. They feed and rest along the way, taking as long as two months for the journey.

Once in a while, especially when disturbed by humans, a swallow will get trapped between stacks of crates or in some narrow spot in our barn. When this happens I do my best to rescue the bird and in so doing perhaps get the chance to hold it briefly, to look into the dark, frightened eyes and feel the small clutching feet and the rapid beat of a tiny heart. Cupped in my hand, weighing less than an ounce, the bird does not struggle to free itself, perhaps sensing I mean no harm, or else aware of the futility of resistance against such overwhelming power. But the instant I step outside and open my hand, it is gone like a leaf on the wind.

One year, long ago, my favorite cat, a calico named Huntress, climbed up on a ledge in the barn and swatted a swallow out of the air right in front of my eyes. I was not pleased and immediately grabbed it from her and set it loose outside. It flew awkwardly for a hundred feet or so, then disappeared in some tall grass. I did not see that swallow again and soon afterward its mate disappeared. That year there were no baby swallows born on the farm. It was in the early days, soon after we had purchased the land, when often just one pair of swallows came each season.

I like to think that our organic farming practices and open-door policy have made the farm more hospitable to swallows, and indeed to all birds. Fields sprayed with chemicals to kill weeds and pests are surely less inviting to insect-eating birds, especially when they are parents with hungry mouths to feed.

Every year there comes a time in late August, perhaps while we are walking back to the toolshed at the end of the day or loading the truck for market, when one of us notices that the swallows are absent from the sky. Some ancient stirring or memory has told them they must set out on their journey south; we, who stay behind, are reminded that the days are getting shorter and that the plenitude of summer will soon be gone. In their departure we recognize the progress of the seasons and the ebb and flow of living things.

That these small birds should come back to our farm each year to hunt the summer skies and raise their young in the barn amid hanging sheaves of garlic and human activity—that they should still do this, even as we humans rush to undo their world (and ours), teaches me how much there is to lose.

ORGANIC CERTIFICATION
and the United States Department of Agriculture

t was with some trepidation that I and many other organic farmers awaited the USDA's December 2000 pronouncement on organic agriculture. After more than a decade of false starts, delays, inadequate funding, public comment, and rewrites, we were offered the "final rule."

An eighteen-month adjustment and adoption period was granted and briefly extended. Then, in the fall of 2002, Uncle Sam took ownership and control of the word *organic*—at least when it is used to describe agricultural products and the means by which they are produced. Unsanctioned use of the word by a farmer can now bring penalties as high as $10,000. With federal backing, it is assumed that public confidence in the organic label will increase and that greater domestic and export opportunities will open up for large organic farms and agribusinesses, many of which are currently involved in conventional industrial agriculture.

The modern organic movement gained momentum in the rebellious counterculture mood of the late 1960s. For many of the early practitioners, it represented a "back to the land" lifestyle choice and a rejection of the increasing chemicalization and industrialization of farming. But what was for many years a fringe movement, mocked and derided by conventional agriculture, had, by the year 2001, become a $7-billion-a-year industry with an annual growth rate of around 20 percent. Across the nation, small nonorganic farms continue to disappear at an alarming rate, but organic farms are on the increase. This growth surge has been driven by consumer demand and a nagging sense among Americans that all is not well in our food system.

Until the feds stepped in, organic farming rules and standards were determined by more than forty private groups and several state agencies across the nation. For the most part, the system worked reasonably well. Over the years an enormous amount of work was done to develop these standards, most of it by dedicated volunteers who were themselves farmers. Organic farming was a grassroots undertaking, receiving very little support from the larger agricultural community.

Enter the USDA, which found itself in the unenviable position of having to define and regulate a type of farming that, I suspect, it would have preferred to see go away. Organic farming is not well suited to the current corporate agricultural model, nor does it generate much income for the agrochemical industry (which makes pesticides and fertilizers), the feedlot operators (who rely heavily on antibiotics, appetite stimulants, and other drugs), the meat-packing and milk- and cheese-processing giants, or the manufacturers of heavy agricultural equipment. All of these groups have a loud voice when it comes to deciding how farming will be conducted in America. In many ways, they *are* the USDA.

Across the nation, small nonorganic farms continue to disappear at an alarming rate, but organic farms are on the increase.

Federal organic standards pose a problem for the agricultural establishment because they have the effect of further legitimizing organic farming. In so doing, they cast a shadow over the other, more dominant kind of farming—the kind of farming that is built on monoculture and chemicals and feedlots. While this "other" kind of farming still provides the vast bulk of our food, it is increasingly viewed as unsustainable and environmentally hazardous. Conventional agriculture is threatened by the public's enthusiasm for the organic approach. It's not the way they want to do business. But the public's voice is strong, and the marketplace is even stronger. Clearly, organic farming is here to stay.

In December 1997, the USDA issued its first draft of national organic standards for public comment. It seemed to many that these standards were written almost exclusively to accommodate agribusiness and that agribusiness, having noticed that there was money being made off the organic label, had exerted considerable pressure in the corridors of power.

The 1997 draft was very permissive and would have greatly weakened the requirements for organic certification. Most glaringly, it allowed the use of genetically modified organisms (GMOs), sewage sludge, and irradiation of food—all anathema to the organic community. Organic farmers and the public were outraged. The USDA received more than 250,000 letters and comments condemning the proposed standards, more negative comment than for any other program in the agency's history. Virtually no mail was received in support of it. It is not surprising that the USDA took a very serious second look. No one likes that much egg on their face.

What the USDA came up with in December 2000 was definitely an improvement. It was, in fact, almost a 180-degree turn. Instead of easing restrictions on what organic farmers could use, the USDA, in some cases, tightened them. The "big three"— GMOs, sewage sludge, and irradiation—were prohibited. The USDA finally listened to the organic community and the consuming public, and then went even further. The 2000 rules were strict on the use of organic seeds and transplants, which at that point in time were not readily available to farmers. Manure and even farm-produced compost could be used only if they met certain criteria. The restrictive nature of several of the new rules has impacted small farmers and deterred some from seeking certification, but most of us have adjusted to them.

There are more serious questions in terms of costs. Many of the old certifying groups, such as the Northeast Organic Farming Association (NOFA), continue to function as certifiers but are now regulated and accredited by the USDA. The USDA, in effect, certifies the certifiers. The federal government's involvement adds a layer of bureaucracy and cost to the process that has not been greeted with enthusiasm.

A great deal of taxpayer money continues to be used to underwrite conventional agriculture, and much public money is used to offset soil erosion and remove pesticides from our soils and water. We might question why our government does not encourage and support a shift in practices that would likely result in a more sustainable, healthy, and ultimately less costly food system.

It is important to note that organic certification has never been a lighthearted matter or merely an expression of good intent. At least it hasn't been since I first got certified by NOFA-NY in 1988. The NOFA-NY standards manual, which is revised each year, has more than a hundred pages of rules and procedures. The annual application that farmers like myself fill out consists of at least fifteen pages. It calls for a detailed farming plan for the season ahead: what crops will be grown and where, and a list of all inputs that the farmer might use (composts, natural fertilizers, botanical and biological pest controls, and so on). Crop rotation, cover cropping, and soil conservation measures must be described. During the year the farmer is required to keep a detailed record of all inputs and all harvested crops and must be able to trace a specific crop sold on the market back to a specific field and the inputs, if any, that were applied.

Each farm is subjected to an annual on-site inspection by a trained inspector. Depending on the size and complexity of the farm, the inspection might take anywhere from two to eight hours.

Conventional agriculture is threatened by the public's enthusiasm for the organic approach.

It involves an inspection of each field and the buildings and equipment that are used in the farming operation. It also includes a detailed review of all required record keeping and all receipts for purchased inputs.

In addition to examining the finer details of a farm operation, the inspector looks at the big picture and tries to ascertain whether or not the farmer has a system in place that is viable and capable of producing organic food. Is there an adequate workforce and suitable equipment for the scale of the operation? Are there inconsistencies, such as weed-free or insect-free fields, that might bring into question the farmer's organic authenticity?

In my case, the certification process usually takes about two and a half days: two days to fill out the application and get the needed materials together and a half day for the inspection. In the last couple of years the annual cost to me has been around $1,500.

I have never resented this time and cost; I have always felt that organic certification is a good investment both for me and for the land that I farm. And the process of getting certified has definitely helped me become a better farmer. That it took place at a relatively local level always seemed fitting and right.

But now, with the heavy hand of the federal government at the helm, my attitude has changed a little. I fear that the added bureaucracy and cost and the inevitable depersonalization that lie ahead may favor organic agribusiness and take some of the steam and spirit out of what I do for a living.

A GOOD KNIFE

A thin strip of steel curving down to a point, honed on one side to a cutting edge. A simple handle to grip. Perhaps a sheath. Nothing more.

A good knife is an elemental thing, the most basic of human tools, and slightly menacing. Yet in the knowing hand its potential uses are almost infinite. Some people are ill at ease with a sharp knife; some are cavalier and unguarded. Others are made more whole, as though a lost limb were restored to them.

If I were cast alone into the wilderness (at times a not altogether unappealing prospect) and could carry with me just one tool, it would undoubtedly be a knife. With just bare hands my prospects for survival would be dim.

Once I attempted to write a poem about a particular knife I carried with me when I lived in the Australian outback. It rested in a sheath that hung loosely from my belt. Even when it was not needed I would often reach to feel its reassuring presence. As hard as I tried, I could not find the right words to describe its functional simplicity and its value to me. I ended up with only an expression of attachment:

After so many a crooked mile
As we've jogged along together,
I'll not trade you in
For another.

I never did trade in that knife. Wouldn't have dreamed of it. But it no longer jogs along beside me. Today it rests in a drawer in my bedroom. Once in a while I take it out and hold it and test its edge and remember the long days we spent together in the wilderness.

For me, one of the many satisfactions of being a farmer is that I can legitimately own and use knives on a daily basis. I have many. A lifetime of knives surrounds me—even lifetimes other than my own, since I have knives that belonged to my father and to my wife's grandfather, and to others long gone. Each is imbued with its own personal history and memories of being held and used by human hands. Among my prized knives are an old bone-handled Bedouin dagger I picked up in Jordan and a Masai machete with a stretched-skin sheath from the Kenyan highlands.

There are also newer, more recently acquired knives. Today I seldom go outside without a Leatherman on my belt—not so much a knife as a multipurpose utility tool, a step up from the ubiquitous and versatile Swiss Army knife. The Leatherman has needle-nose pliers and a wire cutter, as well as screwdrivers, a file, the obligatory can and bottle opener, and, of course, a sharp, workmanlike blade.

*Knife Rack on
Pickup Truck*

It is well made and sturdy enough to be truly useful around the farm. Not having to run back to the toolshed for a pair of pliers or a Phillips-head screwdriver saves time and boot leather. But most often it is the knife blade that is called to use.

For harvesting vegetables and herbs we rely on larger, rigid (non-folding) knives with stainless steel blades. Most of them are purchased secondhand from a neighbor who recycles knives formerly rented to restaurants, hotels, and caterers. Over the years I've probably bought a hundred of them at discount prices. Some have a lot less steel than they did when they started out in life, but they are solid, serviceable knives and their steel is of high enough quality to hold an edge through many cutting strokes. Their blade lengths range from five to ten inches; my preference is for a wide ten-incher.

A Good Knife

73

Our work is easier when our knives are sharp. And safer, too. People respect sharp knives and expend less effort in using them. A blunt knife will cut through cheese and spread butter, but it's not good for cutting heads of broccoli or the woody stems of sage plants. Knife-sharpening is a skill not easily mastered. We use several different stones and steels to maintain an edge, and periodically I employ the services of a professional sharpener.

Learning to use a knife well in a particular application takes time, practice, and intent. Not all individuals are suited to the work. The eye, hand, and arm, and even the whole body should come together in the act of cutting. Of particular importance are the angle of the knife in relation to the object to be cut and the focus of the cutting stroke. When the blade is sharp and the body understands the motion, the effort is minimal and the outcome satisfying. When any element in this equation is missing, the results are usually painful to watch—at worst, strenuous and undignified sawing, hacking, and striking. Wretched to behold.

Lettuce-Harvesting Knife

Each of my workers chooses his or her own knife at the beginning of the season. I encourage them to become familiar with the feel of the knife they have chosen, its weight and size, its balance, the different parts of the blade and what the knife can and cannot do. I ask them to keep it clean and sharp, accord it a certain degree of respect, and always know its whereabouts. When not in use, the knives are usually kept in a knife rack on the side of our farm truck or on a shelf in the barn milk room.

Most of us keep rubber bands on the handles of our knives and follow a defined procedure for cutting and bunching herbs. The right-handed person, after cutting a bunch of basil, for example, will tuck the knife under his or her left arm, peel a rubber band off the protruding handle, wrap it around the stems of the basil, throw the bunch into a crate, and then reach for the knife again. In this sequence the armpit serves as a temporary holster. A variation on this approach uses the left thumb, rather than the knife handle, as a repository for rubber bands. In either case, a skilled buncher, working in a good stand of basil, should harvest a hundred bunches an hour. On an average picking day in the fall, when the harvest is at its peak, we might cut sixty crates of greens and more than five hundred bunches of herbs.

For harvesting lettuces, a special lettuce-cutting knife is used. Instead of getting narrower toward the end of the blade, this knife gets wider and is broadly rounded or angled toward the tip. The user simply presses the end of the blade into the stem of the lettuce a fraction of an inch above ground level. The move is a downward thrust rather than a stroke.

Finger-cutting accidents are most likely to occur when someone grabs a knife of a different weight, size, or sharpness than the one he or she is accustomed to and thus misjudges a cutting stroke. I insist that everyone wear a glove at least on their noncutting hand—the hand that reaches for the plant to be harvested. I keep a collection of old gloves with nicked index fingers and put them on display at the beginning of the season to make my point. However, my warning and instruction are not always heeded, in which case the inevitable bloody, throbbing finger is an effective teacher.

I've always admired the poem "Cutting up an Ox," written in the fourth century BCE by the renowned Chinese poet, Chuang Tzu. The poem describes a certain Prince Wen Hui's cook in the act of preparing an ox for his master's table. The task is accomplished with such immaculate timing, grace, and harmony that the poet likens it to a sacred dance performed to the murmuring sound of the blade.

On the farm, I doubt we will ever achieve the sublime state of Prince Wen Hui's cook as he butchered his master's ox; such are the exigencies of getting the job done in a hurried and imperfect world. Still, I take pleasure in working with a good knife and, once in a while—when the sun is low in the sky and a light breeze carries the mingled aromas of bunched sage, thyme, and rosemary—once in a very great while, I believe I have heard the blade of my knife softly humming as it goes about its work.

The eye, hand, and arm, and even the whole body should come together in the act of cutting.

IN PRAISE
of HERBS

*G**rowing up in* New Zealand in the years after World War II, life was relatively rosy. The troops were home from Europe and Africa and the Pacific. The Allies had triumphed after a long struggle and some sense of rightness and order and possibility for humankind was restored. There was a generous, open spirit across the young land. After school, we kicked rugby balls on neighborhood streets or went sledding on the steep, grassy slopes of nearby hills, with a little brown paper bag of mutton grease to speed our descent.

In the summer holidays we went swimming at nearby beaches or hitched rides down to Wellington harbor to fish for herring and mackerel. Sometimes, to my delight, we visited my uncle's thousand-acre sheep farm in the Wairarapa—a farming region north and east of Wellington that in my youth was primarily a home to several million sheep; nowadays the Wairarapa still has its share of those woolly locusts, but it also boasts numerous vineyards, apple orchards, and olive groves, and even some organic vegetable farms.

Every child had enough to eat, or so it seemed, and almost every New Zealander could read and write. The worlds of church and state were comfortably distinct from one another. The elected governments and many of the people (there were about two million when I was a child) had mildly socialistic inclinations. The idea of commonwealth, of sharing what the nation had to offer, was current back then.

We had good food to eat. New Zealand was and still is an impressive agricultural producer. The soils are fairly fertile and the climate eminently suitable for growing a wide variety of crops and the year-round grazing of animals. We ate meat (mostly lamb or mutton) and fish and plenty of fresh vegetables and fruits, milk, cheese, bread, and butter. There was no shortage of English-style cakes, cookies, and desserts, which have left me with an unfortunately sweet tooth to this day. It was simple fare, without a lot of frills and flourishes.

Most herbs and spices, if known at all, were viewed with some suspicion as being foreign or European in origin and not sufficiently British. The word "herb" was pronounced with a hard *h* and invariably used as an abbreviation for the male first name Herbert.

Salt and pepper were the condiments of choice. A little vinegar on fish-and-chips or on a plate of freshly sliced tomatoes went down rather well. My mother did use a fair amount of the curly parsley that she grew in her garden and also made the occasional curried dish, no doubt a carryover from England's long colonial involvement in India. I remember being especially partial to hard-boiled curried eggs.

Such herbs as sage, rosemary, thyme, basil, oregano, and dill were hardly in the New Zealand vocabulary sixty years ago, but all that has changed, and so have I.

Today New Zealand is more global in flavor, more like the rest of the world. Where there once were fish-and-chips shops and homemade meat pies, now the blight of McDonald's and Kentucky Fried Chicken is on the land. But there are good things, too. The food is still mostly fresh and healthy and moderately priced. New Zealand's wine industry has come of age, and fine restaurants serve international cuisine with a local flavor. Herbs from around the world are commonplace now. My sister, Claire, grows garlic and several other herbs in her garden and uses them freely.

And I, too, have become a grower of herbs—approximately twenty in total, if you count our half-dozen varieties of mint and two or three basils and thymes. People come to our farm stand from all over Manhattan and even the outer boroughs of the city to buy the fresh herbs and garlic that we offer. Sometimes they ask my advice. Which herb or combination of herbs would be right with a particular dish? I struggle to be helpful and may refer to a printed, laminated sheet that I assembled, titled "Herbs and Their Uses." But in truth, though I know my many herbs quite well from a grower's point of view and certainly appreciate them on several levels, I am not a reliable source for culinary information.

It is perhaps natural, given my background, that the first herb I attempted to grow was parsley. In the early years, before our green-house was built, I seeded parsley directly in the field along with everything else and soon discovered that this is one plant that takes its sweet time to germinate. Folklore has it that parsley goes to the devil seven times before breaking ground in this world of ours and will still decline to sprout if sown by a dishonest man.

Some twenty-five or thirty days after I sowed it, the first tiny parsley plants emerged amid a sea of weeds. With newfound respect for my mother's gardening prowess, I settled down on hands and knees for several hours to create a clean environment for this, my first herb. The parsley begrudgingly repaid my efforts on its behalf by growing very slowly for the next few weeks, but then it took off and started moving with impressive speed. At last I felt vindicated, and before long I was taking bunches of curly parsley to market.

They sold well enough, but my customers were quick to tell me that they would have preferred flat—Italian—parsley.

Today we grow both curly and Italian parsley, but always twice as much Italian—in New York, it is the parsley of choice, at least for cooking. Curly parsley is more often used as a garnish, though I myself have no qualms about eating it. Last year some fair amount of our parsley was trimmed down by a groundhog before I took action against the beast. Interestingly, this parsley-eating marmot, like my New York customers, showed a distinct preference for the Italian variety.

I now seed most of my parsley in the greenhouse and have learned that by soaking the seed in water for a day or two prior to planting, I can cut the germination time in half, eliminating a couple of those possibly quite interesting visits with the devil.

The second herb I attempted to grow was basil. I was strongly encouraged to do this by my wife, who assured me that New Yorkers would flock to the stand in large numbers if I offered them homegrown tomatoes and fresh basil in the summer months. Of course, she was right. Basil turned out to be easier to grow than parsley, and every bit as appealing to my customers. The seeds germinated quickly and the young plants flourished in the summer heat. But I soon learned that basil cannot handle cold weather. Frost will kill it outright. This means that our season for selling basil lasts barely a hundred days, whereas we can have parsley at the stand for a solid five months.

Seed catalogs offer many varieties of basil, and I've tried growing quite a few of them—Lemon-scented Basil, Cinnamon Basil, Purple Ruffles, Dark Opal. They sound wonderful and, in fact, each is exceptional in its own way; but, as is the case with many unusual vegetables and herbs, the exotic varieties often lack broad, continuous appeal. These days we stick with the basics—Genovese and Italian Large Leaf. They are the best basils for making pesto or for serving with fresh tomatoes. And they are what the majority of my customers want.

If you exclude garlic, which we place in a category of its own, parsley and basil are the foundation and mainstay of our herb business. We plant at least a couple of thousand of each every year and

Just before
Thanksgiving,
when the
stuffing of
turkeys is a
national
pastime, sales
of rosemary,
sage, and
thyme can
leap into the
stratosphere.

hope for multiple cuttings. The three next most important herbs on our farm are sage, rosemary, and thyme, no reference to Scarborough Fair intended. These popular Mediterranean herbs prefer hot and moderately dry conditions. All are perennials, which means they are capable of living and remaining productive for several years. Some of my sage plants are a decade old and I feel almost as though I know them personally, having wrapped my hands around them many a time and harvested their fragrant limbs for market.

Thyme, perhaps fittingly, is more sensitive to the passage of time and has less longevity in our humid and often wet climate. Rosemary is a tender perennial and cannot tolerate severe cold. While it may function as a long-term ground cover in southern California, in the Northeast it will invariably die if left outside in the winter. In the last few years I've overcome this limitation by overwintering a fair amount of rosemary in our high tunnel. This plastic hoop house, though lacking an artificial heat source, stays noticeably warmer during the day—and a few degrees warmer at night—than the open fields. It also stays a lot drier. When planted in the tunnel and mulched with straw or wood shavings and then covered with a transparent, breathable blanket, my rosemary has survived some harsh winters and come into the spring looking quite jaunty. This pleases me greatly. I'm very fond of rosemary, and especially of its neat appearance and tangy, almost piney aroma. And in the winter months, there's nothing quite like a cup of fresh ginger and rosemary tea.

Demand for rosemary, sage, and thyme is at its peak in the fall, when my customers are turning more serious attention to the kitchen and the preparation of hot and substantial dishes. But it's important to have these staple culinary and aromatic herbs at the stand all season long. This way we establish ourselves as serious herb people. While we may sell just twenty bunches of rosemary or thyme on a Saturday in June, by mid-October or November the number can jump to fifty or sixty of each, per market. And, just before Thanksgiving, when the stuffing of turkeys is a national pastime, sales of rosemary, sage, and thyme can leap into the stratosphere. On the Wednesday before Thanksgiving, our sage sales have been known to exceed four hundred bunches.

By midsummer, oregano, marjoram, French tarragon, summer and winter savory, and lavender can all be found on display at our stand. Many customers have their favorites and know exactly what they are looking for, but there are others who are less confident and less sure of their herbal needs, and to these poor souls we are willing to throw out a little offhand advice now and then. Oregano, of course, is recommended to those looking toward a pasta dish, especially if it involves tomatoes. Marjoram, I've learned—probably from more knowledgeable customers—is excellent with carrots. French tarragon is a strong, almost licorice-flavored herb that can transform a chicken dish but should not be used in excess. The savories—both summer and winter—taste like a peppery thyme. I like to tell my more formal-looking customers that either savory will go well with dishes involving beans and that, by the way, should it be a matter of concern to them, the savories will lessen the bouts of flatulence that often follow such meals.

Lovage is an unusual perennial herb somewhat akin to celery but with a distinct aroma. While a few chefs who buy from us have found good uses for this uncommon herb (which my computer's spell-checker does not recognize), most of my customers have no idea what to do with it. I say most, but definitely not all! We've discovered an important niche market for lovage in the form of Romanians. They simply crave the stuff. If you grew up in Romania, it seems, you'll always feel a little malnourished and misused by life if your diet is lacking in lovage. It is regarded as indispensable to the hearty meat, vegetable, and bean soups that Romanians enjoy in the winter months. The lovage is usually chopped up, like parsley, and added to the pot ten or fifteen minutes before the soup is finished.

Somewhere in the borough of Queens, in New York City, there is a small but thriving Romanian community, and within that community the word is out that lovage can be found at our stand. Almost every Wednesday and Saturday at least a few Romanians make the pilgrimage into Manhattan in search of their much-loved lovage.

The word is out that lovage can be found at our stand.

They often pick up ten or fifteen bunches but rarely buy anything else from us. They've told me that they freeze most of what they buy for future use. Invariably, they are delighted to take what lovage we have. If, however, they arrive at the stand to discover we are sold out, they can be painfully disappointed. So popular is this herb among Romanians (I'm told Poles, Bulgarians, and a few others from that region of the world have a soft spot for it, too) that on days when we have a good supply on hand, we'll tape up a large sign proclaiming WE HAVE LEUSTEAN—the Romanian word for lovage. This always brings a warm smile to the faces of those travelers from Queens.

Herbs stand apart from the many vegetables we grow. Alone they are not quite foods, yet with their essential flavors and fragrances, they also are more than food.

And, of course, for our English-speaking customers, the word *lovage* lends itself nicely to evocative and humorous promotion. ALL YOU NEED IS LOVAGE sometimes appears on our signs and garners a laugh now and then; and, this past year, Melanie, one of my interns, came up with: HAVE YOU BEEN LOOKING FOR LOVAGE IN ALL THE WRONG PLACES? OURS IS FRESH, ORGANIC, AND COSTS ONLY $1.50 PER BUNCH.

On the farm, our herbs are scattered around in at least a half-dozen different locations. Annuals such as basil or cilantro and biennials such as parsley (which we treat as an annual—planting a new crop every season) are rotated through the fields and often planted in black plastic, with drip irrigation lines laid underneath.

Perennial herbs may remain in one spot for several years. If they like plenty of moisture, as do our chives, catnip, lovage, lemon balm, and several mints, we plant them on lower ground and mulch them heavily with composted wood shavings. The organic mulch preserves soil moisture and helps suppress weeds. Those perennial herbs that prefer dry conditions, including sage, rosemary, and thyme, are usually planted on high ground and into a sturdy weed-barrier fabric. Laying the fabric and stapling it to the soil, then making holes in it for the herbs, is more work up front but will result in less weed-pulling in the years to follow.

Every few years we set up a new perennial herb patch on the farm and begin phasing out one of the older herb gardens. By moving our herbs around in this manner, we hope to avoid the buildup of diseases and depletion of soil nutrients that might result from the constant cutting of one herb in the same place year after year. Herbs are not all alike in their growth habits and needs: Some have deep roots and some shallow. Some are content to keep their roots in one place; others migrate laterally (mints are particularly good at this). And each herb has its own nutritional needs. If an herb stays in one place too long and is taxed by repeated harvest for market, it may run short on those soil nutrients that are critical to its success. As though cognizant of this, many of the perennial herbs are quite amenable to being dug up by the roots on a damp overcast day in spring, divided into several pieces, and replanted elsewhere.

Herbs stand apart from the many vegetables we grow. Alone they are not quite foods, yet with their essential flavors and fragrances, they also are more than food. By themselves, they may not sustain us, but in conjunction with the more common staples of life, fresh herbs bring nuances of taste and zest to our diets that add immeasurably to the pleasure of eating. Without them, our lives would be poorer.

Many herbs have strong health-giving and curative properties and have been used through the ages to good effect. "The spice of life," which surely encompasses the entire world of herbs, is a far from hollow expression. Each herb is a plucky and spirited individual with a wild streak—always a little beyond the control and manipulation of humans. Most are strong and independent survivors whose pedigree extends well beyond our own.

Herbs have several attributes that go beyond the culinary and curative and further commend them to a grower like myself. While often slow to start, once established, they are strong and persistent growers. They are resistant to most plant diseases and generally unappealing to vegetarian insects (the Japanese beetle's fondness for basil is one notable exception).

Most herbs have developed good defenses against animals such as groundhogs, rabbits, and deer, which allows us the option of planting them in the less protected and farther reaches of the farm. From time to time, we lose a little parsley to these nonhuman residents, and last year, some denizen of the hedgerows developed a taste for new shoots of basil, but the great majority of our herbs arrive at market entirely unmolested. The same cannot be said for many of the vegetables we grow. Our lettuces and other greens are constantly under assault. More than a few times we've offered special deals on "partial lettuces" with signs that read: PRE-EATEN or FEEL CLOSER TO NATURE—SHARE YOUR LETTUCE WITH A WHITE-TAILED DEER. Some of our customers are willing to indulge us, but not many.

Herbs are light and easy to carry and have relatively high value. Fifty bunches of rosemary, weighing only a few ounces each and selling for $1.50 a bunch, will bring in a pleasantly unstrenuous $75. Compare this with a sixty-pound crate of pumpkins or butternut squash, which will tax the muscles and joints but might net a mere $50 return. (The older I get, the more I appreciate this ratio of weight to retail value. The point becomes even more salient when one considers that the average sixty-pound crate of squash may be picked up and moved by human hands (ours) a half-dozen times before point of sale: Upon harvesting, the squash are packed into a crate in the field, then carried to a truck or wagon, then transported to the barn where they are unloaded and stacked. On the day before market they are carried from the barn and loaded onto a different truck, from which, at a later point, they must be unloaded and displayed at market. Those that do not sell must be carried home, unloaded, reloaded, and so on. From the carrier's point of view, that sixty pounds of squash can easily turn into something more like three or four hundred pounds of effort. The few pounds that the average crate of herbs will weigh precludes any concerns of this sort.)

At the stand, herbs take up obligingly little display space. Their unique aromas are a treat for our customers, even those who choose to go home with no herbs in their shopping bag. With a favorable breeze in the air, basil just released from the cooler can draw passersby from twenty feet away. In the fall, fresh rosemary and thyme seem to set off just the right sensory receptors for gastronomic activity. While most people visit our stand for more basic foods, many will pick up an herb or two at the end of their purchase to complement and enhance the meals they have in mind.

With a favorable breeze in the air, basil just released from the cooler can draw passersby from twenty feet away.

Herbs are pleasant to work with. Unless rain threatens, we cut and bunch our herbs in the late afternoon when the sun is less intense and the hard work of the picking day is mostly done. Herb cutting, unlike much of our other harvesting, can be quiet and contemplative—a time to allow nature to enter our human sphere, a time to feel the gentle western wind on our daylong faces, a time to listen to birdsong, to notice a chipmunk or squirrel on the edge of a field or a hawk with an eye for detail circling overhead. It is perhaps even a time, if we are able, to soften the sharp edges of our lives and accept what fate and fortune have dealt us.

To those who might be drawn to the challenges and rigors of the truck-farming life and to urban markets and the possibilities they present, I would strongly urge the inclusion of at least a few herbs in the crop plan. They should win you new customers, line your pockets with extra dollars, and, by the example they set, strengthen your resolve when it is weak and lift your spirits if they should flag.

FARM DOGS

*W*e *now have* three dogs. Kuri, my first dog and indisputably the top dog on the farm, is probably a mix of coonhound and Doberman, but no one knows for sure. His legs are long and tawny brown. His coat is black, with symmetrical brown markings on his chest and face. His body is lean, angular, and muscular. His ears are floppy and his eyes have a fiery gleam in them.

But perhaps his most notable feature is his tail, which is robust, unusually long, somewhat crooked, and highly active. During bouts of enthusiastic wagging, this talesome tail has been known to knock small children off their feet.

Kuri

Kuri's origin is a mystery that, I suspect, will never be solved. He showed up on our front lawn, as a full-grown dog, one spring morning in 1989. In those days it was not uncommon for neighbors' dogs or strays to pass through the farm, but none of them ever had the effrontery to settle down in the middle of the lawn barely ten feet from our house. With the righteous indignation of a new property owner, I shouted at him to be on his way and, to my satisfaction, off he went with his substantial tail hanging low between his legs.

But a half hour later, when I stepped outside to fetch something, there he was again. This time, perhaps in a show of deference to me, he was sitting at least twenty feet from the front door. I yelled some curt words at him and he promptly ran off once more.

An hour later he was back, sitting a little farther away, looking straight at me. This time, I hesitated a moment and returned his gaze, and in so doing I saw something in his wild, plaintive eyes that softened my resolve to be rid of him. I also noticed that he was very thin, almost emaciated, and it occurred to me then that he might be hungry.

I went back inside, put some cat food in a bowl, and brought it to him. As I approached he backed away, no doubt unsure of my intentions. I set the bowl on the grass and stepped aside. Kuri then got down low, almost onto his belly, and in a slinking, submissive fashion, made his way to the bowl of food. When he arrived at it he gulped down the contents in five seconds flat. I got more food and brought it to him. With his oversize tail wagging furiously from side to side, he licked the bowl clean a second time, then looked up at me, pulled back his lips, and wrinkled his nose, almost as if to smile. A moment later he turned and trotted off in a light prancing gait, with his tail high in the air.

Later that day I noticed he was resting in my tractor shed and I dimly understood that, in spite of myself, I had made a new friend. Kuri (his name, which he received a few weeks later, means "dog" in the Maori language of my native New Zealand) has been here ever since. The farm is his place as much as it is mine. He knows every nook and cranny of it and every woodchuck hole. Together we have watched the seasons come and go and together we have grown older, only he more so than I. Perhaps I flatter myself, but I have come to believe that Kuri is as fond of me as I am of him, that he would defend me, even to the death, if it came down to that.

Aldo and Tiki, our two other dogs, are brother and sister, nephew and niece of Bruzzi, the handsome beast and murderer of chickens who died a couple of years ago. Like their uncle, they are purebred Maremmas. Originally from Italy, especially the mountainous Abruzzi region, the Maremma's role in life was to protect sheep from wolves and other predators. They are large, white, bear-like dogs with thick coats—fast, powerful, independent, and (some people think) slightly dangerous. They are also quite beautiful and have a playful innocence about them.

All three dogs are working dogs. They live outdoors year-round. For Aldo and Tiki, this is no hardship—they are loath to enter a doghouse, preferring the shelter of a tree or an open lean-to, even in the coldest and most inclement weather. They delight in play-fighting and rolling about in the snow and will lie down in it for hours. In the summer heat they are less content and will seek out shaded areas where they dig large holes to settle in. Their talent for excavating has given portions of our lawn the appearance of a minefield despite the great displeasure I have expressed over this behavior.

Kuri is in partial retirement. He is a short-haired dog and an old one, with a bit of arthritis setting in. The cold winds of winter are no joy to him. If invited, I'm sure he would move into our house in a minute (and perhaps he will be invited one day). But I fear it would be the beginning of the end for him, that the easy warmth and comfort of the indoors would rob him of his vitality and spirit and that he would submit to the inevitable decline that awaits us all.

A few years ago I installed a heat mat under a false floor in his doghouse. He was highly suspicious at first—on cold nights, when the mat was turned on, he would stand outside his house and shiver in protest. But eventually he accepted the new technology and now seems the happier for it.

Together we have grown older, only he more so than I.

Aldo and Tiki are less than two years old. My wife still refers to them as "the puppies," though they are more like rambunctious teenagers. Tiki is faster and more nimble than Aldo, and she is also a better hunter. Both usually eat what they kill, be it a rabbit, a rat, or a woodchuck, but Aldo, who is bigger and has bigger jaws, likes to eat what Tiki kills if he can prevail over her. This leads to angry clashes with much growling and gnashing of teeth. Interestingly, when it comes to dry dog food, their proprietary instincts disappear—they will happily eat from the same bowl, nudging each other's nose out of the way without the slightest sign of hostility.

Like all Maremmas, Aldo and Tiki have acute hearing and are especially alert at night when wolves might be on the prowl. The slightest noise—a rustling among the leaves, a twig cracking in the distance, a field mouse passing wind after a heavy meal—will set them off on a barking frenzy. There are no sheep on our farm to be threatened by wolves and, alas, no wolves, but there are vegetables that are constantly under threat from the ever-increasing population of deer. Aldo and Tiki's job is to keep the deer away, though they do not know this. To them, deer are to be eaten and vegetables are in the somewhat irrelevant category of all plant matter, hardly worth protecting, even with a single bark.

Lacking sheep, their instinct is to protect us and the house in which we live. We appreciate their concern for our safety, but do not feel endangered. Therefore, as darkness falls and deer begin coming out of the woods to feed, Aldo and Tiki are led to the fields and chained near those vegetables that are most vulnerable to assault. Because the deer make some sound as they move about and probably leave a scent in the air, the dogs bark at them, and this, in conjunction with a single strand of electric fence and the occasional shotgun blast, is usually enough to spare our crops.

One of our neighbors has taken more direct action to control his deer problem. He obtained a special permit from the state to shoot deer out of season when they are eating his pumpkins. I don't know how many he killed this summer, but he gave me three carcasses over a period of a few weeks. One of them was fresh and I butchered it for our own use. The others were somewhat aged when I got to them and a little tainted by the summer heat—not ideal for human consumption, but just right for dogs.

Spoils of the Road—Tiki

To all three dogs, the gutting and butchering of a deer is a major event and one which, as spectators, they participate in with great attentiveness. If I were not present they would gladly do the work themselves, in their own fashion, and within a week a full-size deer would be reduced to a few stray bones and some unpalatable entrails. Under my supervision, the job is done in a more orderly manner: The meat and bones are cut into meal-size portions, wrapped and frozen, to be parceled out over time. But always, a few fresh pieces—a chunk of liver or heart, the fat from around the kidneys—are given to each dog while the work is under way.

The first time Aldo and Tiki were present when I butchered a road-killed deer (picked up on the way home from market), they attempted, with great enthusiasm, to assist me in my efforts. It was not a workable situation. Between my knives, saws, and axes, and their teeth, something was bound to go wrong. I put leashes on them and tied them to a fence within viewing distance. Minutes later, Aldo, then just six months old, had chewed through his leash and was back trying to chew through the deer's hind leg. It was then that I learned the inadequacy of ropes and leashes, at least where Maremmas are concerned. To restrain these dogs, it is necessary to use chains.

Now we have arrived at a certain dining etiquette.

Now we have arrived at a certain dining etiquette. I have established myself as the alpha male and, during the dismemberment of a deer carcass, the three dogs, unrestrained and with eager anticipation, watch me from several feet away, glad to be there and to receive the generous morsels I throw to them. For all of us, it is a bonding experience.

MARRIAGE
of BODY
and MIND

F arming, more than most occupations, relies on the use of the
body as well as the mind—and, even more so, on the fruitful
marriage of the two. These days, physical work is often devalued and
disdained, better left for someone else to do. Not so in farming, or at
least in the type of farming I practice. Small-scale organic agriculture
values physical work and the lean, efficient use of the body. It draws
upon the whole person. For this reason it is enlivening.

In our modern, high-tech world we are so invested in the brain and
its cleverness that we often forget we are also physical beings and that
our bodies appreciate some healthy, productive outdoor use. There

is a sort of primal satisfaction that comes from a day of work in the fields. It has more pith and substance to it than a jog in the park or a workout at the gym. But farming should never be romanticized—it is real work. Those with an idyllic notion of farm life will be quickly disabused if they ever encounter the reality.

Nor is the physical labor required on a farm simple, unreasoning work. There's more to using even a rake, a hoe, or a shovel than most people understand. Newcomers to these tools are invariably stiff and uncomfortable. They depend heavily on their arms and shoulders and quickly tire themselves out. The more seasoned shoveler, to use but one example, puts to use the legs and torso and knows how to use the body as a fulcrum and the eyes to direct the body's movement. Such an individual has learned how to breathe deeply and evenly and pace him- or herself. He or she can shovel for hours, if need be, with an almost effortless, rhythmic elegance.

The same is true of almost every task we perform, from planting to weeding to raking to mulching to picking to lifting and carrying crates of produce. The person with experience, the one who has considered the task and how the body might best move through it, is at a tremendous advantage and will greatly outperform the novice. But more than just experience is required. A person's openness and attitude are most relevant. Those who regard physical work as demeaning, simplistic, or beneath their dignity are not likely to be much good at it.

It is gratifying to witness the transformations that take place in many of my young workers over the course of a season. In the beginning, when their bodies are unaccustomed to the demands being placed upon them, they may feel inadequate or uncertain. They are likely to experience discomfort, exhaustion, frustration, and perhaps some mental anguish and resistance.

One young man who has since developed into a committed and talented farmer told me that in his early days he had to deal with frequent internal temper tantrums. When confronted with a seemingly endless weeding job, voices inside his head would start telling him, "This is ridiculous. I'm on my hands and knees pulling weeds. I have a college degree. I wasn't intended for this." And perhaps even more vexing, all the while that he was being bombarded by these negative thoughts and emotions, there were others, working alongside him, who seemed to be moving along happily enough. He didn't want to be seen as a quitter, so he stuck with it. Eventually the job would be done and he would feel a surge of relief and some sense of accomplishment. He could see that his efforts had made a difference.

Now this individual is at home in the field and more at ease with his body. He enjoys physical work and is willing to take on the challenge of almost any job. When negative thoughts intrude, he doesn't entertain them for very long.

As we approach the end of a season, some of my interns tell me how difficult it was for them in the beginning: the bending, the sun, the discovery and conditioning of muscles they didn't even know they had, the almost nonstop, seventeen-hour market day that left their brains fried and their bodies barely able to transport them into bed. But they laugh as they talk of these early challenges and hardships and go on to speak of how good they now feel. They've begun to take pride in their work and the results that are there for everyone to see—in the form of multiple rows of glistening vegetables and herbs. They also speak of being more in tune with their body's needs—the need, for example, for a good, protein-rich breakfast, or a substantial and carefully prepared evening meal, or a midday nap. And at night they enjoy their rest and feel they have earned it.

In our modern, high-tech world we are so invested in the brain and its cleverness that we often forget we are also physical beings.

Looking at these young workers, it is obvious that their bodies are stronger and fitter (and frequently leaner) than when they arrived, and that they have discovered within themselves an easy physical confidence. They no longer view a particularly daunting job with loathing or trepidation. They go to it, achieve what they can, and then move on to the next one. If they are fortunate, they have learned it is possible to marry their mental and physical selves to good effect.

At this point in history, our society tends to elevate and reward the specialist, the individual who has directed most of his or her energy and time toward the mastery of one skill or discipline. This concentrated focus has brought some benefits, among them a certain kind of affluence and an unending supply of goods and services. It has fueled our consumptive, growth-obsessed, capitalist economy. It is the modern way, the modern creed. It may also be a modern malady. Specialization, when taken too far and allowed to define who and what we are, becomes limiting. It robs us of our wholeness and our self-sufficiency. It misses the big picture and confines us to a narrow zoom. And it leaves us at the mercy of experts.

For me, small-scale farming is a remedy for our modern condition. It is one of the few human endeavors left that favors the generalist over the specialist. Indeed it *requires* a generalist. The university-educated agricultural specialist may find a home at ConAgra or DuPont or in academia, but he or she will not necessarily perform well on a small organic farm. An altogether different mind-set is needed.

IT'S a LONG ROAD to a TOMATO

That fresh, locally grown, vine-ripened tomato, surely one of summer's most relished gifts, has its inception in the mind of the farmer in the short, cold days of winter. Outside, the land is frozen and most likely blanketed with snow. Inside, the body is warm and rested from the season that has passed and the brain is ready to entertain new and bounteous possibilities.

Seed catalogs begin arriving in the mail: Johnny's Selected Seeds, Fedco, High Mowing Seeds, and a half dozen others. Glossy photos of immaculate vegetables and fruits are a feast for the winter eyes, and the accompanying descriptions of flavor and yield are equally tantalizing.

Tomato

Of course, in the catalogs little mention is made of the time and the toil required to bring such beauties to fruition; nor does the eager mind dwell on such irksome details.

Every year there are new varieties to consider as well as old standbys that always make the list. First in line are the heirlooms and open pollinated varieties: Brandywine, Yellow Brandywine, Cherokee Purple, Black Krim—the names themselves call forth an enticing range of visual and gastronomic experience. Then come the cherries: Super Sweet 100, Sungold, and Sun Sugar—each one sweeter than the one before. The list is rounded out with a couple of varieties of plums and a few favored hybrids such as Celebrity and Early Girl. Eventually, orders are placed and, for several weeks, no additional effort is required.

In late March, after the onions and perennial herbs have emerged in the greenhouse, it's time to start the first batch of tomatoes. We seed tomatoes on three separate occasions—a few weeks apart, and about a thousand each time. Our goal is to transplant into the field about 2,500 plants over a period of six weeks. This strategy spreads our risk and ensures a staggered yield and longer period of harvest. The little, yellow-brown tomato seeds, seemingly desiccated and lifeless, are sown in plastic flats containing potting mix. Our seeding medium is a compost-based mix from McEnroe Organic Farm in Dutchess County, New York. We use at least a ton of it every year. The flats we use contain 162 cells—each about 1 inch square and 1.5 inches deep. A handful of potting soil is sprinkled on top of the seeds.

After seeding, the flats are misted with water and placed on rubber electric heat mats that provide heat from below and keep the soil temperature at around seventy-five degrees. If kept moist, the seeds will germinate in about eight days. It's always a small pleasure and relief to see the first ones appear—to witness, in our uncertain world, nature's constancy and endless desire for regeneration.

Once they are established, we thin the seedlings to one per cell. Adequate warmth, daily watering, and perhaps the occasional foliar spray of fish emulsion and kelp, for nutritional health, are all that is needed. A month later, when the plants are standing three or four inches tall, they are moved, or "potted on," to flats with much larger cells. Here they reside for a few more weeks, with fresh potting soil and ample room to spread their roots. After a short period of adjustment they quickly put on new growth.

In mid-May, we take the seedlings from the controlled and sheltered environment of the greenhouse to the larger, more demanding outdoor world. They are set outside on something we call a "hardening-off structure," which provides some shade from direct sun but otherwise exposes them to the elements. Four or five days later the young tomato plants are ready for the big move to the field.

Well in advance of field planting, the land must be readied. Preparations usually include a winter cover crop of oats or rye, the spreading of compost, and then spring tillage. This last step calls for a tractor and chisel plow to uproot any weeds or other unwanted plant matter, and then a rototiller to incorporate all organic material into the soil and create a level bed. Such work should be performed under suitable soil moisture conditions—preferably not too dry and definitely not too wet. Often the weather does not cooperate—this season, five weeks of nearly continuous rain in May and June (totaling more than eleven inches!) severely tried our patience and greatly hampered spring planting.

We plant about two thirds of our tomatoes in four-foot-wide strips of black plastic and the other third in aged wood shavings from a nearby horse farm. Both are considered mulches; their purpose is to block weeds and retain soil moisture. The plastic is cheap and effective but must be disposed of at the end of each season, an unpleasant and guilt-ridden task. The wood shavings are not as good at blocking weeds, which means more labor for us,

When we have done our job well . . . we will know from our customers' enthusiastic remarks.

but, as a side benefit, they add organic matter to the soil as they gradually decompose.

Before the black plastic is laid on the soil, lines of drip-irrigation tape are run down the center of each bed. These enable us to irrigate the tomatoes under the plastic, precisely in the area of the plants' roots. This practice conserves a great deal of water. Drip irrigation also keeps the plant foliage dry (in contrast to overhead sprinkler irrigation), which reduces the incidence and spread of plant diseases. Holes are made by hand in the plastic and the tomatoes are planted twenty inches apart in a row on one side of the drip tape. As a small parting gesture, each plant is usually given a shot of water, a teaspoonful of rock phosphate, and an equal amount of an organic 2-4-2 fertilizer. These amendments are given as insurance and are probably not necessary—the soil, when managed well, should meet most, if not all, of the plants' nutritional needs.

For a couple of weeks after transplanting, the young plants look quite small and forlorn in the large expanse of the open field. It is as though they are taken aback by the scope and conditions of the new world to which they have been consigned. The wind dries out their tender leaves, the rain pelts them without mercy, the nights in late May and June can be chilly, and the midday sun reflecting off the black plastic can be uncomfortably intense. Woodchucks, rabbits, and deer may be tempted to take a few bites before deciding that tomato foliage is not to their liking. It's all quite a shock after the easy life in the greenhouse and the short stint on the hardening-off structure.

In time, most tomatoes make it through (the few that don't are quickly replaced) and soon find the great outdoors is conducive to their growth and well-being. Within a month they have doubled, then tripled and quadrupled in size. The hot sun is good to them; the soil is alive, deep, and hospitable. Twice a week, unless rain falls, the plants receive a couple of hours of water via the drip-irrigation lines. Flowers soon appear, and, shortly afterward, the long-awaited fruits develop.

When the plants are twelve or fifteen inches tall, we set up a support system to keep them growing upright rather than sprawling on

The hot sun is good to them; the soil is alive, deep, and hospitable.

SAY AHH

A N ESSAY ON tomato production on our farm would not be complete without some discussion of pronunciation. Of the many Britishisms acquired during my formative years in New Zealand, almost all have fallen away or been replaced by solid American language and usage. Not so with the word tomato. I retain the pronunciation of my youth—"toe-maah-toe"—emphasizing and lengthening the second syllable, for two good reasons.

First, it is not right for a man to shed all of his heritage when adapting to a new land. Second, the lengthening of the middle syllable into an "ah" sound is more satisfying and, I believe, more expressive of the true nature of the subject. I will concede that the clipped American pronunciation—"ta-may-toe"—is perhaps quite appropriate when referring to the bland supermarket tomato given to us by agribusiness. But, when speaking of tomatoes sold at our stand, and the stands of other small local growers, I firmly believe it is proper and right to dwell on that second syllable. The long "ah" sound captures the deeply satisfying flavor of a really good, honest tomato.

the ground. This entails pounding several hundred four-foot stakes into the soil every six feet or so, down the center of each row of tomatoes. Supporting twine is then run between the stakes. As the plants grow taller, more levels of twine are added. This system of trellising is known as the Florida weave, not to be confused with the more lively Florida jig (popular among the state's many retirees) or the nefarious Florida shuffle (first performed on election night 2000).

Under the right conditions, tomatoes, like most of us, will prosper and grow plump and sweet. They like warm nights and hot sunny days with temperatures in the 80s; they like rich, loose soil with lots of organic matter. They like just the right amount of water, especially when setting and developing fruit—generally the equivalent of an inch to an inch and a half of rainfall a week will keep them happy. Too much water causes the fruit to crack and robs them of their sweetness. Not enough slows their growth.

Also, like us, tomatoes can fall victim to various diseases and natural assaults: Early blight, blossom end rot, anthracnose, bacterial spot, and sunscald are just some of the maladies that may afflict them. If disease strikes, there's not much an organic grower can do except hope that the whole crop will not be affected. A healthy soil, warm sun, staggered plantings in different locations, good air circulation, and not too much rain are our best defenses.

As you might imagine, the first harvest of ripe fruit is eagerly anticipated by crew and customers alike. The Sungold cherries are usually the first to delight our palates. Then the Taxis (a yellow, low-acid variety), followed by Early Girl, a small but pleasingly sweet specimen. By the time the heirloom varieties ripen, we're in the heart of tomato season and the plants should be laden with sumptuous, richly colored fruits of various shapes and sizes. Sweetness and flavor are at their peak. The picking is easy. The market is bustling. And the return on investment can be quite good. But, as with so many good things, fresh, local tomatoes do not last very long. In the Northeast, the season for these epicurean treasures, circumscribed by frost at either end, is lamentably short—if we get six or eight weeks of respectable harvest, we are satisfied. Ten weeks makes us ecstatic.

As with so many good things, fresh, local tomatoes do not last very long.

When we have done our job well, when nature has blessed our endeavors, we will know from our customers' enthusiastic remarks and the broad smiles on their faces. Their friendly support and patronage is the final and most satisfying measure of our success.

THE PRICE of MILK

(December 2002)

E ddy Bennett is a dairy farmer and neighbor of mine. Forty-
four years ago he was born on the land he farms today, as was
his mother some forty years before that. Every morning of every
day, Eddy gets up to milk his herd of Holstein cows. He is a shy but
friendly man with a ready smile and a keen sense of humor. Milking
cows is the only work he has ever known—and he knows it well.

Eddy tells me that when he was a boy almost everyone in town had
cows; he can name at least a dozen families and points in different
directions toward the land they once farmed. Today, there are just three
operating dairies left in the Orange County town of Greenville,

where Eddy and I both live. These three probably won't be able to hold on much longer. Chronically low milk prices and rising overhead costs have taken a heavy toll; suburban sprawl plus an indifferent and largely unappreciative public have not exactly boosted morale.

Small dairy farmers, like Eddy, tend to be self-reliant and conservative. They are accustomed to making their own decisions. Often they work alone or within the framework of a family. As a breed, they are not temperamentally suited to self-promotion, lobbying, or collective bargaining. In this regard, they are at a considerable disadvantage: The forces they must contend with—middlemen and retailers; a national milk-pricing system that favors certain regions of the country; fluctuating supply in the marketplace; soft demand; and government policy and regulation (it's illegal for Eddy to sell me a quart of milk on the farm without a special permit, for example)—are overwhelming.

Once a nation of farmers, most of us today are far removed from our agricultural roots and have scant grasp of what it takes to put the food we've come to expect on our tables. The other day, while bemoaning the crunch small dairy farmers find themselves in, I got a fairly pat response from a friend: "Don't the cows provide the milk?" my friend asked me. "How difficult can that be? Everyone I know feels overworked and underpaid. At least farmers have land. That's more than I've got."

This little tirade got me thinking, and I decided to ask Eddy Bennett just what it takes to bring us a seventy-five-cent quart of milk. We sat down at his kitchen table one morning and he gave me the hard facts and the hard numbers. Both were harder than I expected them to be.

These days Eddy milks around seventy cows, which is more than he used to. He also looks after fifty heifers and thirty calves. Rickety fences and a patchwork of cropland, pasture, and woods define the 190-acre farm he inherited from his parents. The house he lives in was built more than a hundred years ago. The road he lives on, Kurpick Road, carries his grandfather's name.

Eddy has one employee—Kevin Miedema, the seventeen-year-old

It's a Long Road
to a Tomato
104

son of an ex-dairy-farming neighbor—who helps him after school, on weekends, and in the summer. The rest of the time, Eddy's on his own. He works a seven-day week, usually 15 or 16 hours a day, fifty-two weeks a year. That averages out to about 108 hours of work per week. I asked Eddy if he ever takes a vacation. He flashed me a sardonic grin and replied, "I've had three. All of them in the hospital: two for hernias and one for blood poisoning."

THE SCHEDULE

DAIRY COWS NEED to be milked twice a day—just about everything Eddy does, including sleep, revolves around this recurring necessity. An average day in his life looks something like this:

7:30–8:30 AM: Get up and have breakfast.
8:30–10:30 AM: Feed heifers in barn.
10:30 AM–1:00 PM: Milk cows, then turn them out to pasture. Clean barn.
1:00–1:30 PM: Eat lunch.
1:30–6:00 PM: Do field work (plant, mow or bale hay, harvest corn, fix fence, spread manure, etc.).
6:00–8:00 PM: Bring cows back to barn and feed them.
8:00 PM–midnight: Eat and sleep.
Midnight–4:00 AM: Milk cows and feed calves. Clean barn.
4:00–7:30 AM: Sleep.

Squeezed into this long day are many other chores, such as fixing and maintaining equipment, ordering supplies, and keeping records for the accountant.

Repair and maintenance of equipment is an ongoing challenge, almost a part-time job in itself. To keep his farm functioning, Eddy relies on five tractors and a formidable array of mechanical and electrical equipment. Many of his implements are old and prone to malfunction. Almost every day something needs fixing; when a breakdown occurs, Eddy must diagnose the problem, obtain the parts, and make the necessary repairs. The cost of hiring others to do this work, except when absolutely necessary, would be prohibitive.

Cows, especially when they are giving milk, have big appetites, and it's up to Eddy to keep them well supplied with corn, hay, and supplemental feed. To this end, he grows fifty acres of corn each year and cuts around two hundred acres of hay. To produce the corn and hay, Eddy needs the equipment listed below. The prices in parentheses represent the approximate cost of each item were he to purchase it new (though he almost never does—which is why things break down a lot).

THE EQUIPMENT

Corn-Planting Equipment:
   Plow, to turn over soil ($8,000)
   Disc, to level soil for seeding ($8,000)
   Planter ($15,000)

Corn-Harvesting Equipment:
   Chopper, to cut and chop corn and throw it into a wagon ($40,000)
   Wagons, to transport corn (two @ $10,000 each)
   Silo blower, to blow corn into silo ($4,000)
   Silo unloader, to retrieve corn when it has become silage ($7,000)

A different set of implements is needed for making hay.

Hay-Making Equipment:
   Mower and crusher ($18,000)
   Hay rake, to form the hay into rows ($4,000)
   Tedder, to fluff up the hay so it will dry faster ($4,000)
   Square baler ($20,000)
   Round baler ($15,000)
   Bale wrapper, to wrap round bales ($15,000)

Hay wagons, to transport square bales (two @ $2,500 each)
Loader bucket, to pick up round bales ($7,000)

To operate this equipment, Eddy sometimes uses three of his five tractors simultaneously. When making hay, for example, one tractor might pull the round baler, another a hay wagon, and a third might be used to lift the bales onto the wagon. Each tractor, if purchased new, would cost around $50,000. At any given time, because they are old and put to frequent and heavy use, one or two of Eddy's five tractors may be out of commission.

Other essentials include a barn ($100,000), a silo ($20,000), milking equipment ($20,000), a barn cleaner ($8,000), and a manure spreader ($12,000), all of which must be maintained in working order.

For those who are inclined to add numbers, the total estimated value of all the above equipment, if it were purchased new, is $600,000. Because most of Eddy's equipment is far from new, and there is not much demand these days, in our area, for barns and silos, the actual value is probably under $200,000. Then, of course, there are the cows. Depending on age and fecundity, each animal in Eddy's herd is worth between $500 and $1,200.

FARM INCOME

CURRENTLY, EDDY IS receiving around $12.50 for every hundred pounds of milk his cows produce (100 pounds of milk is equivalent to 11.6 gallons). This price, $12.50 per hundred pounds, is about where milk prices were twenty-five years ago. Unfortunately, the price of virtually everything needed to run the farm has either doubled or tripled over that same period of time. Just the price of milk is the same—the same for Eddy, that is; not for you and me. If we could go back in time twenty-five years to a supermarket or grocery store, we would pay around 40 cents for the same quart of milk that costs 75 cents today.

A dairy cow usually lives for seven years. If successfully bred, a cow will start producing milk at two years of age. Each of Eddy's cows gives an average of 17,000 pounds (almost 2,000 gallons) per year. The cows are milked for ten months a year, then get two months off. His herd of seventy milking cows should produce a total of 1,190,000 pounds (138,000 gallons) of milk in a year.

At the going rate of $12.50 for one hundred pounds of milk, Eddy's gross annual milk sales would be approximately $149,000. But because milk prices were higher earlier in the year, his gross milk sales for 2002 will probably be closer to $155,000. Sales of cows, calves, and heifers might bring another $4,000, and he should receive $2,000 in tax credits and refunds.

Eddy also will receive about $8,000 in agricultural price support payments, through a program called MILC (Milk Income Loss Contract). When milk prices drop below $16.94/hundredweight (a price the federal government deems adequate for farmers to make a living), the government chips in approximately 45 percent of the difference between $16.94 and the prevailing market price. This would bring Eddy's estimated total gross income up to $169,000, which seems quite impressive—until one looks at the expense side of the ledger.

FARM EXPENSES

EDDY'S SINGLE BIGGEST expense is purchased supplemental feed, which his cows consume at a rate of five tons per week. This pelletized soy-corn-mineral mix costs $50,000 a year. Farm supplies account for $24,000 and labor another $20,000. The list goes on from there.

Approximate annual expenses on Eddy Bennett's farm:

Feed	$50,000
Supplies	$24,000
Depreciation	$23,000
Labor	$20,000
Spray and fertilizer	$5,000
Trucking	$5,000

Gas/Fuel	$2,000
Insurance	$6,000
Employee FICA and medical	$1,500
Rentals and leases	$5,000
Repairs and maintenance	$6,000
Utilities	$5,500
Veterinary fees	$5,000
Accountant	$2,000
Advertising/Promotion/Marketing	$1,000
Other Expenses	$4,500
TOTAL EXPENSES	$165,500

Eddy is not a man who spends money unwisely. There is very little wiggle room in his expense budget. Most of the expenses listed above are set by forces beyond his control. Some, such as advertising and promotion, are charged to him whether he likes it or not.

The Bottom Line

Summary of annual income and expenses on Eddy Bennett's farm, fiscal year 2002:

Milk Sales	$155,000
Livestock Sales	$4,000
Other Income	$10,000
TOTAL GROSS INCOME	$169,000
TOTAL EXPENSES	$165,500
NET PROFIT	$3,500

It doesn't take an accountant, or the proverbial rocket scientist, to see that Eddy Bennett, and small- to medium-size dairy farmers like him all over the Northeast, probably will not stay in business much longer, under prevailing prices, no matter how hard they may work, or how committed they may be to their land

and the life they lead. This means that a huge part of our history and landscape will disappear—will give way to more housing developments and strip malls.

This is happening even though demand for milk in southern New York State far exceeds local supply. Industrial megafarms, mostly in the Midwest and California, will be all that is left to provide our nation with cheap milk. Of course, once all the smaller farms are shaken out and the industry consolidated, milk prices will probably go up.

A Different Scenario

IF THE PRICE OF milk in our supermarkets went from 75 to 80 cents per quart and the extra nickel was earmarked for local farmers in proportion to the amount of milk they produce, Eddy's net profit would go up to about $31,000. He and many farmers like him might find that quite acceptable.

THE HIDDEN COST of FARMING

O*nions, 79 cents* a pound. Broccoli, $1.49 per head. Carrots, five pounds for $2.99. Potatoes, ten pounds for $3.99. I found these prices while cruising the aisles of my local supermarket last week. Pretty good, I'd say. But a little scary: there's no way I could match such prices and stay in business.

There are not many places in the world today where fresh, raw food is so plentiful and costs so little relative to income as in the United States. "Supermarket to the World," boasts one of our American agribusiness giants, and with some justification. For most of the twentieth century (and now into the twenty-first), high production and low cost have been the cornerstones of American agricultural policies.

It's hard to argue against inexpensive food—I like a bargain myself. But before we get too self-congratulatory about our food system, we should take a closer look. Behind it lie some hidden and largely unacknowledged costs that can make our cheap food not such a bargain after all.

Consider the cost of pesticide use. Every year close to a billion pounds of pesticides (herbicides, insecticides, fungicides, anti-microbials)—about twelve hundred in total, with names like Ambush, Bladex, Roundup, and Pounce—are sprayed across American fields. Yet the percentage of crops lost to pests has gone up, not down.

These pesticides (poisons, really) accumulate in the soil, in our food, and in our bodies. They leach into the groundwater and run off into streams, rivers, and estuaries. The Environmental Protection Agency has recognized agricultural pesticides as a serious health hazard. Many that are now banned as unsafe enjoyed years of widespread use. A report released by the Centers for Disease Control in 2005 indicates that the average American carries about a dozen pesticides in his or her body. The typical pesticide body burden includes chemicals that have been linked to cancer, hormone disruption, and neurological and birth defects. Children and migrant farm workers are the most likely to be adversely affected.

We spend plenty of money trying to clean up our waterways and restore fish and wildlife habitat. And we spend plenty more removing pesticides, nitrates, and other farm pathogens from our drinking water. It's impossible to say how much pesticides add to the health-care budget, but add they certainly do. These are hidden costs we seldom take into account when we bite into that cosmetically perfect apple.

On privately owned cropland in the United States, close to two billion tons of soil are lost to erosion each year due to intensive farming practices and heavy use of agricultural chemicals (over fifty million tons of fertilizer are used annually). The poisoned and lifeless topsoil gets washed into our rivers and blows away in the wind.

The Natural Resources Conservation Service (NRCS), a federal agency, and equivalent state and county agencies (all fifty-seven counties in New York have a Soil and Water Conservation District), spend huge amounts of money trying to reduce soil erosion and repair other environmental damage. The NRCS budget for 2009 was $3.47 billion. For 2011, President Obama has requested $3.99 billion. Meanwhile, the soil keeps disappearing.

In Orange County, New York, where I farm, a third of the cropland is eroding at a rate of more than six tons of topsoil per acre per year; on some slopes the rate can exceed ten tons per acre per year. This is way beyond the rate of soil replacement by natural processes, such as the weathering of rocks and the decomposition of plant and animal remains.

Money spent attempting to slow the rate of soil loss, while probably well spent, is another hidden cost that doesn't show up at the supermarket checkout counter. But it surely is buried somewhere in our tax bills, which fund the work of the state and federal remediation agencies. When viewed in this light, the much-touted efficiency of American agriculture can ring a little hollow. Efficient for whom, we might ask.

Not long ago, we decided that cigarette manufacturers should pay some share of the health-care costs associated with the use of their products. Perhaps it's time to assess chemical manufacturers for the environmental damage their products cause, or at least tax those products in a manner that reflects the cost of cleanup after their use. As it stands now, we, the taxpayers, pick up the tab for chemical-dependent agriculture: We get cheap food, yes, but we pay for it later, while the agri-giants who dominate much of American farming get richer and more powerful by the day.

If we are serious about promoting sustainable agriculture, why not *reward* farmers who look after their land and treat it as a truly renewable resource. Offer them financial incentives to reduce soil erosion and chemical use—it would save us money in the long run. We should encourage the use of cover crops, green manures, and crop rotation. We should give tax credits to farmers who let a portion of

There's no way I could match such prices and stay in business.

their land lie fallow for one year in five. We should reward diversity rather than monoculture. These are farming practices that have worked for centuries and are the bedrock of a healthy food system.

We hear a lot these days about the disappearance of the family farm from the American landscape, and we lament this loss. The truth is that low food prices and subsidized overproduction move small farms closer to bankruptcy and accelerate the trend toward consolidation of our food system into the hands of megafarms and feedlots.

Some farmers put off replacing the old tractor or repairing the leaky barn roof. Some try to milk more cows, working around the clock to do it. Some call a real estate agent and start looking for another line of work. In the five years between 2002 and 2007, New York lost 903 farms. Most of them were dairies. In the last several years the rate of loss has accelerated. Is this what we want? Is that 99-cent quart of milk really such a bargain after all?

Low food prices and subsidized overproduction move small farms closer to bankruptcy.

In 1999, a study completed at the University of Essex in England came up with some startling figures. It estimated that farming costs British taxpayers 2.3 billion pounds annually over and above what they pay for their food. This amount is almost equivalent to the total income that British farmers receive. The 2.3 billion pounds goes to cleaning up polluted water and air, repairing habitat, and covering the cost of food poisoning.

Here in the United States, Cornell researcher David Pimentel came up with similar findings. In his paper titled, "Environmental and Economic Costs of the Application of Pesticides Primarily in the United States," he estimates that we pay an annual $10 billion price tag for environmental and health costs associated with pesticide use. At the human level, Pimentel has calculated that pesticide use results in 10,000 to 15,000 cases of cancer, annually, and 300,000 incidents of food poisoning. Environmental impacts include the poisoning of wildlife, especially birds, and other natural enemies of crop pests.

Of course we can't single out modern agriculture as the sole culprit for all our environmental problems. Our high-consumption, high-waste, growth-obsessed industrial/technological society must share plenty of the blame.

But, speaking as a small farmer, I know we could do things differently. Rather than continue to spend huge amounts of money ameliorating the effects of modern factory farms and agribusiness, why not start investing in a more earth-friendly approach?

We could create a commodity price structure that doesn't discriminate against small farmers. Most small farmers want to be good stewards of the land, but our nation's agricultural policies and artificially low prices have not encouraged stewardship. Instead they have encouraged intensive cropping practices and a heavy dependence on chemicals to boost production, both of which degrade the land over time and consolidate its ownership in the hands of a few.

Let's stop subsidizing the agribusiness spoilers and polluters who view farmland merely as a commodity and whose lobbyists buy favors in Washington with huge campaign donations. Let's put research money into organic and sustainable systems, not new chemicals and genetically engineered plants.

The overwhelming majority of people would like to see more small farms, more local production, less chemical use, fewer feedlots. We must convince our government and the industrial agribusiness complex that currently holds the reins that we want and expect something better.

When you pay $2.99 for five pounds of carrots, it may feel like you're getting a bargain. But it's a phantom bargain, not a real one. Under our current system of farming, you'll be paying the balance due at a later date—on your tax bill or at your doctor's office.

WINTER WORK

W*hat do you* do in the winter, Keith?" I'm asked at least a couple of dozen times a year—usually in late November or December, as our marketing season draws to a close and we advise customers that we won't be around much longer. Every year, a few days before Christmas, we decamp from our farmers' market and don't return till late May or early June. That's a solid five months we're not around. Those who buy at our stand on a regular basis are understandably curious about what goes on during that time.

Do I retreat to some Caribbean island for an extended bout of sun and sand therapy? Take on another line of work? Sit back and read a few good books? Or do I slip into a state of suspended animation and allow the world to go on its way without me?

Some assume I hop the first plane to New Zealand and take advantage of the antipodean summer. Back when my parents were still alive, my wife and I did make trips to New Zealand, but only once every three years or so, and never for more than three or four weeks. That still left a lot of time to account for.

So what does a farmer like me do in the winter? Well, lots of things, actually, including some of the aforementioned. But, if there's time to give an inquisitive customer a comprehensive answer, I generally begin with a description of some basic farmerly activity, such as the ordering of seeds and supplies. This is done through catalogs and occasionally over the Internet and is a more challenging task than one might think. It can easily take a week or more, and for good reason.

Placing orders in the winter is a critical early step in planning for the season ahead. There are many factors to consider: How much land will be available in the field-rotation scheme for each crop? How many workers will I have and what will their level of experience be? What equipment is available to assist us in our efforts and is it in good working condition? How much energy do I, personally, expect to have (of late, I've noticed a decrease each year)? And then there is the critical question of demand: How much of each crop can I reasonably expect to sell?

A lot of this is guesswork, but it is educated guesswork based on experience and lessons learned in past years. In the deep of winter, seed catalogs can be very seductive. They speak of bountiful possibilities, but don't dwell on the multiple efforts required to transform seeds into saleable vegetables. Nor do they touch upon the many things that can, and often do, go wrong. In farming, setting your sights too high is a dangerous habit. You must resist the tendency to overestimate your productive capacity while seated in a comfortable chair.

If I order too much of a particular item, especially something perishable like potatoes or shallot sets (with these crops, actual potatoes and shallots are planted, rather than seeds), I'll almost

You must resist the tendency to overestimate your productive capacity while seated in a comfortable chair.

certainly want to use them. This means more labor will be required—to plant, weed, water, mulch, harvest, store, and sell that crop. Since there's a finite number of hours in a day, there will naturally be less time available for other crops, which may end up suffering from neglect. Now, if I've grown more potatoes or shallots than I can sell in a season and, in the process, shortchanged other crops for which there is unmet demand, I've used my time unwisely and will suffer the consequences.

As soon as seeds and supplies arrive, I begin organizing the greenhouse—cleaning off benches, laying out heat mats, obtaining the necessary potting mix and other materials. By mid-February it's time to start seeding onions and perennial herbs. Each week thereafter has its own seeding schedule: root crops like celeriac and kohlrabi; lettuces and assorted greens; tomatoes and peppers, basil, summer squash, zucchini, and so on. By mid-April there will be approximately 150 flats containing about twenty-five thousand plants in the greenhouse. From the beginning these need daily attention—in the form of watering, ventilation, and heat. If one of these requirements is not met, trouble will soon be on the way. Once the greenhouse is in operation, trips to the Caribbean or the Land of the Long White Cloud (a translation of Aotearoa, the Maori name for New Zealand) are no longer an option.

Winter is also a time for prodigious amounts of paperwork. Each year an application to the Greenmarket must be filled out and submitted. This multipage document asks for, among other things, a comprehensive crop plan—i.e., a list of all vegetable varieties to be grown, quantities to be planted, planting and harvest dates, and anticipated harvest amounts. For me, the Greenmarket crop plan usually runs to about eight pages.

The annual application for organic certification is also many pages long. It calls for a fairly detailed description of the whole farm operation, including a field plan or chart, showing the location of crops, and all inputs and methods used. Two or three days must be set aside to complete it. Like the ordering of seeds and supplies, these two applications are part of the planning process. They force

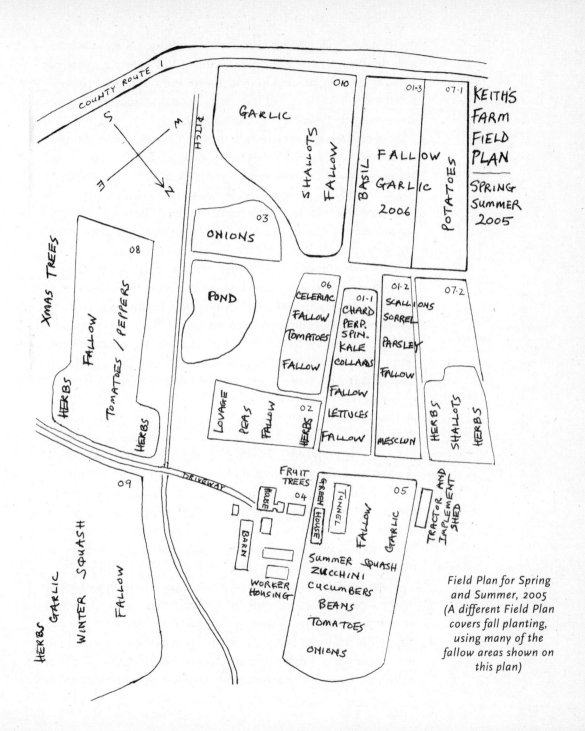

COUNTY ROUTE 1

S
E
W
N

DITCH

GARLIC

SHALLOTS FALLOW

O10

BASIL
FALLOW
GARLIC
2006

O1·3   07·1

POTATOES

KEITH'S
FARM
FIELD
PLAN

SPRING
SUMMER
2005

03

ONIONS

POND

XMAS TREES

08

FALLOW

TOMATOES / PEPPERS

HERBS

HERBS

06
CELERIAC
FALLOW
TOMATOES

FALLOW

O1·1
CHARD
PERP.
SPIN.
KALE
COLLARS

FALLOW

O1·2
SCALLIONS
SORREL
PARSLEY

FALLOW

07·2

HERBS SHALLOTS HERBS

02
LOVAGE PEAS FALLOW HERBS

FALLOW
LETTUCES
FALLOW

MESCLUN

DRIVEWAY

09

HERBS
GARLIC

WINTER SQUASH

FALLOW

FRUIT TREES

HOUSE   04

BARN

WORKER
HOUSING

GREEN HOUSE

TUNNEL

05
FALLOW GARLIC

SUMMER SQUASH
ZUCCHINI
CUCUMBERS
BEANS
TOMATOES
ONIONS

TRACTOR AND
IMPLEMENT
SHED

Field Plan for Spring
and Summer, 2005
(A different Field Plan
covers fall planting,
using many of the
fallow areas shown on
this plan)

me to make choices and decisions well in advance of actually going into the field. Though testily resistant to this type of paperwork, I've come to regard these applications and others like them as unpalatable medicines that impose order on my farming endeavors and very likely benefit me in the long run.

Also on the paperwork front is the onerous task of assembling in a coherent fashion all data necessary for a pre–April 15 visit to the accountant, without whose help and guidance my business might soon descend into some purgatorial chasm. Though I wish it were otherwise, the volumes of paperwork that flow toward me throughout the year are often pushed aside and neglected during the growing season when the demands of vegetables trump most other needs. By default, winter is a major time for paperwork catch-up.

Winter is also the time to hire a crew. Unless I'm lucky enough to have one or two returnees, this means finding a half dozen superworkers. I need people who are bright, fit, motivated, eager to learn, and willing to work long hours for low wages, under sometimes difficult conditions (rain, heat, cold). These rare individuals must also possess social skills that will enable them to live and work together in relative harmony. They must also be trustworthy and capable of performing many tasks, including running the market stand without me. Finding all these attributes combined in one individual is a tall order. Finding six such individuals in a couple of months is about as daunting as crossing a midsize ocean in a rowboat. In an average year I correspond or speak with thirty or forty applicants, interview ten or more in person, and hopefully am able to lure the best of them to our farm. Interviews take four to six hours and sometimes involve lunch, dinner, and an overnight stay. Then references are contacted. It all takes time.

Though perhaps more content with my own company than most people, I do on occasion enjoy a little noncommercial social interaction. Winter is definitely the time for it. During the growing season, my social life is severely restricted by the imperative to sell vegetables and make money. Getting home at nine thirty on a Saturday night after a seventeen-hour day at market pretty much rules out movies, dinner with friends, or the occasional party that might come along. Nor is Friday night any better, since I must be in bed by nine-thirty or ten o'clock in order to get the minimum five hours' sleep I need to see me through the following day. The other nights of the week are not much better. Generally, the goal is to be in bed by ten o'clock and up by six. In winter this schedule is quickly thrown out the window. A hot bath and a glass of sherry or Armagnac at midnight is just fine, so long as there will be no demands placed upon me early in the morning.

At this stage, the inquiring customer, if still present and attentive, is probably satisfied that productive activity takes place in the winter and may have learned more than he or she wants to know. It all gets a bit fuzzy and doesn't leave a distinct impression. I know this because sometimes the same individuals make the mistake of asking the same question in future years. When this happens, assuming I am remembering correctly, I might give a very different answer—one that takes into account a few other aspects of what appears to be a rather large block of free time.

Perhaps I'll tell them, in an offhand manner, that I spend my time sitting by the fire with a cup of tea, pondering the larger issues pertaining to human existence, such as the tyranny of power, the destruction of nature, and the short- and long-term prospects for the continuance of my species. Admittedly, this scenario contains a slight untruth, since we do not have an open fire in the house. But it's likely to elicit a more interesting response than any description of placing seed orders, filling out forms, or visiting an accountant. Some people respond with a sigh and a knowing smile, others with a nervous laugh and a quick change of subject, and still others with a mild look of bemusement, even stupefaction. "Well," they might say, "the important thing is that you come back here with your vegetables in the spring." I agree with them, both in word and in spirit. That is the important thing. Then I remind them that I won't be back till early summer.

In the end, the most important message to leave with my urban customers is this: If they want to eat fresh local vegetables, someone nearby has to grow them. This takes time and a good amount of work. A lettuce destined to be sold in late May must be started in mid-March. The onions seeded in February won't become saleable bulbs till the end of July. Tomatoes eaten in August and September begin their journey to the dinner table in March or April. These most obvious truths are often not well understood. The other day (I believe it was on March 15) a good friend who frequently inquires about and visits the farm asked my wife if I was back selling at market yet. Never mind that the ground was still frozen solid and coated with six inches of snow.

But, to be fair, these days some industrious northern growers are using greenhouses and high tunnels to extend their growing season well into the winter months. They can be found selling baby lettuces and salad greens and sometimes an assortment of sprouts in March and April. A few others go to market all year round, relying on such storage crops as carrots, turnips, parsnips, potatoes, winter squash, onions, and garlic to generate needed income. These robust individuals must stand outside all day long and endure the aching cold of winter and a greatly diminished flow of customers. They are far hardier than I.

A lettuce destined to be sold in late May must be started in mid-March.

When you are separated from the growing process, as most people are, you tend not to think about the labor and time that goes into the creation of local food. Growing vegetables is not an assembly line process completed in a day. Nature takes its time to bring us a potato or a bulb of garlic, and a fair amount of human intervention is needed along the way. While the winter months may afford some time for rest and replenishment, and perhaps even a little philosophical reflection, they are far from idle. Though you probably will not see them, most farmers are still farming in the winter.

GROWING POTATOES

I n my first few years of farming, I shunned potatoes. It's not that I had an aversion to eating them—on the contrary, I've always enjoyed a good spud. It was more a case of vegetable snobbishness. To be honest, I saw potatoes as being too common and too cheap to interest me. Let other farmers grow potatoes and turnips and carrots—the staples of the masses. I'd stick with more delicate and exotic crops, especially those likely to fetch a good price.

As is the case with most snobbery, my attitude was smug, small-minded, and, perhaps even worse for someone operating a farming business, ignorant. Today, I see potatoes very differently. Two things changed my point of view. Over time I noticed, with my retailer's eye, that a few growers at the Greenmarket were selling potatoes for as much as $2.50 per pound. That was very impressively different from the typical supermarket prices of $3.00 or $4.00 for a five-pound bag that I was accustomed to seeing. Admittedly, these high-priced spuds were no ordinary tubers. They carried names like German Butterball, Yukon Gold, and Purple Peruvian Fingerlings. They were, in their own way, exotic. One day I bought a selection from a neighboring farmer and the following night ate them for dinner. They tasted extraordinarily good, each variety subtly different from the others. Their full and deeply satisfying flavors were what turned me around and set me on a more enlightened path. I decided to put potatoes on my crop plan for the following year and give them a try.

A couple of seasons later, it became evident that growing unusual potato varieties is not so easy. More to the point, the yields (the weight of potatoes harvested versus the weight planted) can be quite low—there was a reason my neighbors were asking high prices. After some consideration, I decided to pursue a more modest course. At that time there were not many organic potatoes in the market, and those that did exist were mostly in the exotic, high-priced category. Why not grow a few more-standard, higher-yielding varieties and offer them at prices below those charged for fingerlings and other exotics? This seemed like a good, mid-level niche waiting to be filled. By then I had learned that more ordinary potatoes grown organically on our farm and generally dug the day before sale, while perhaps not quite as good as fingerlings, tasted significantly better than their supermarket counterparts. And my customers seemed to agree.

Now, a dozen or more years later, we are firmly established in the potato business and happy to be there. We plant about 1500 pounds of seed a year, which requires about three-quarters of an acre of land. Our favorite is a Maine potato called Kennebec. It has white flesh, a buff-colored skin, and excellent all-around eating quality. Boil it, bake it, roast it, mash it—Kennebec always comes through tasting good. Another spud I'm partial to is Red Gold. This one has a reddish skin

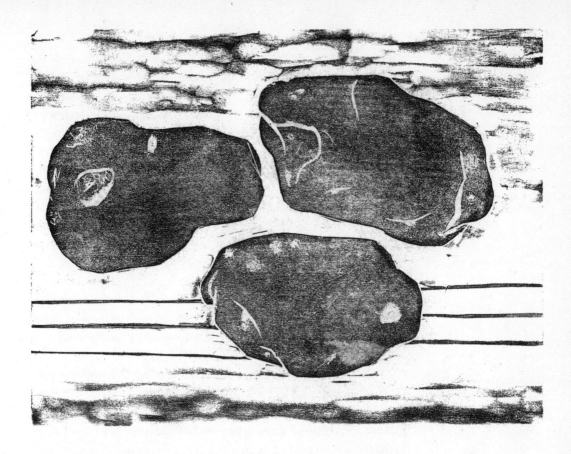

and slightly yellow flesh. The tubers are not as big as the Kennebec ones, but they take less time to reach full size. Red Gold is a good early-season potato. It is excellent steamed or boiled and tossed with a little butter. It also excels in a potato salad.

Another one I like is Carola—a true yellow-fleshed, German potato akin to Yukon Gold. It has a rich creamy texture and superb buttery flavor, which our customers love. But for us, Carola often does not yield well. In dry years, we've harvested as little as two or three pounds for every one pound planted. Kennebec, on the other hand, often gives us six or eight pounds for a pound of seed. We find that quite respectable. Some growers achieve much higher yields (10:1 or even 15:1), but it's usually with the aid of chemicals.

If we have a bountiful harvest, we'll probably have potatoes left over at the end of the year that we may use as planting stock the following season. It's critical that these saved potatoes be disease-free and in good condition; questionable planting stock should be avoided like the plague. One has only to remember the Irish potato famine, which resulted in the death by starvation of as many as a million Irish people and prompted major waves of immigration to the United States. The Irish working poor were heavily dependent on potatoes, and when the fungal disease late blight crossed the Atlantic and struck Ireland in 1845, there was little other food to fall back on. The use of infected planting stock only made matters worse in subsequent years.

. . . it became evident that growing unusual potato varieties is not so easy.

Late blight can still strike fear into the hearts of potato growers. From time to time, pesticide-resistant strains still enter the United States from Mexico or Central America—either on the wind or in the form of infected tubers crossing the border.

Besides late blight, there are several other diseases potatoes can fall victim to. The names are enough to cause concern: blackheart, scab, ring rot, pink eye. None are pleasant. Buying inspected, certified-disease-free seed from a reputable source is the prudent way to go. I've been advised not to use my own saved seed for more than a couple of years in a row since there's no way to be sure that some nasty disease has not crept in undetected. By the same token, it is unwise to plant eating potatoes sold in a local store or supermarket. Absence of clearly noticeable symptoms does not guarantee absence of disease.

When ordering seed potatoes, I invariably go to Moose Tubers, the potato, shallot, and onion division of Fedco Seeds. Fedco is a Maine seed cooperative with a distinctly alternative-political and small-farmer philosophy. The Fedco catalog, while a little daunting at first, is a true work of art presented in black and white on newsprint. No over-the-top glossy photos here, but plenty of beautiful etchings and engravings, quotes from the classics, and offbeat humor, in addition to cultural advice. One recent Fedco catalog pronounced: "The idea that a catalog should do nothing but sell is a fit adjunct to the doctrine that the sole obligation of business is to look to its own bottom line, that maximizing private profit always translates to public gain and therefore that pure capitalism is the best way to solve all social problems. Our catalog aspires to more." (C. R. Lawn)

And more it provides, including an excellent selection of certified-disease-free, organically grown spuds at reasonable prices, made more attractive by Fedco's large-volume discount policy.

We normally plant potatoes at least twice each season—sometimes in late April but more often in early and mid-May. The soil temperature should be above fifty-five degrees and the weather preferably settled. If you plant potatoes at the beginning of a prolonged cool, wet period, there's a fair chance many of them will rot before sprouting new growth. I've made that costly mistake more than once.

In the last few years, prior to planting, we've dunked our seed potatoes in a tub of water containing an organically approved biological fungicide called T-22. Whether this benign and beneficial *Trichoderma* fungus has actually helped, I don't know for sure, but of late our yields have been better and there's been almost no rotting of seed pieces.

I prefer planting whole potatoes (they are less likely to rot), but large ones may be cut into two or even three pieces to make them go further. Each piece should weigh at least two ounces, be no smaller than a good-sized egg, and have a couple of eyes—the little thumbnail indentations from which sprouts emerge.

Planting potatoes can be quite enjoyable. Once the soil is tilled, I ride up and down the field with a tractor cutting pairs of narrow furrows about six inches deep and three feet apart. My crew and I then take plastic shopping baskets full of spuds and walk along the furrows, dropping a seed piece in every ten or twelve inches. We then walk back and gently press the seed pieces in with our feet. Finally, we use rakes to pull a few inches of soil over the top of them.

Depending on the number of people involved (usually five or six), we might plant five hundred pounds in four or five hours. It moves along quickly. Spring is in the air, as is a sense of possibility. The work is active and engages the body in a comfortable way. Once finished, we have the satisfaction of looking out over a fair expanse of land and knowing that an important job has been done. Already, we begin to contemplate the first harvest and the wonderful flavors that await us.

With spring showers to keep the soil moist and enough sun to keep it warm, we'll see the first stubby, dark green leaves break ground in about fifteen days. Under less favorable conditions, they will take longer. After emergence, a little hoeing between and around the plants to control weeds is a good idea. But the young potato plants grow rapidly and are soon able to shade out nearby weeds with their broad and sturdy foliage.

A few weeks after emergence, the plants should have attained a height of eight inches or more. It's now time to hill them. This involves mounding soil around them to the point that only the top few inches of leaf are left exposed. A gardener might do this with a shovel or a rake. We use a tractor and hilling discs and make short work of what would otherwise be a formidable job.

Hilling ensures that the new generation of developing potatoes are covered with plenty of soil and therefore not exposed to sunlight, which causes solanine to develop, creating spuds with a greenish hue. Most of us know that green potatoes are not good to eat—they are, in fact, mildly poisonous and will likely give you a stomachache. Sunlight, penetrating through even an inch of soil, can cause this greening problem. As the plants continue to grow, hilling is performed a second and often a third time. Besides combating solanine buildup, hilling also disrupts weeds that might otherwise establish themselves and compete with the potatoes for moisture and nutrients.

Fertile soil and adequate rainfall will ensure good leaf growth. The more leafy growth the plants are able to achieve, the greater the size and number of potatoes they will produce. The formation of the potatoes—known as "tuber set"—starts to occur about a month after the plants first appear in the field. While the tubers are forming, it's critical that the plants receive plenty of water—ideally the equivalent of one and a half inches of rain each week. If we haven't already set up irrigation, now is the time to do it.

In the life of a potato grower, a year seldom passes in which he or she is not paid a visit by the Colorado potato beetle, known as CPBs to those of us who must confront them. These slow-moving, though winged, black and yellowish-orange striped insects about the size of a coffee bean have a knack for finding us wherever we are. They may overwinter in the soil or come in from elsewhere. Upon arrival, usually in June, they set about laying clusters of orange eggs on the leaves of potato plants. The eggs soon hatch into rather unpleasant-looking pinkish-orange sluglike larvae with black spots along their sides. The increasingly plump larvae have voracious appetites and the capacity, due to their large numbers, to defoliate an entire field of potatoes. One can expect two and sometimes three generations of CPBs in a season.

Conventional growers have several chemical sprays at their disposal for controlling Colorado potato beetles. Organic growers are limited to certain forms of a biological pesticide known as Bt (*Bacillus thuringiensis*). This live bacterium, when ingested by the larvae, causes them to stop feeding and eventually die. Over the years I've used Bt and usually with satisfactory results, especially when the larvae are small (they gain size rapidly as they feast on the plants). But spraying in the summer heat is an unpleasant task, and moreover, the Bt is expensive.

More recently, we've tried a new, low-tech approach: We confront our enemies by hand. It's not a matter of just picking them off individually and squishing them between our fingers or stomping on them—we're more sophisticated than that.

Our method involves the use of shallow plastic tubs about twelve by eighteen inches in size and old broom heads, preferably with fairly loose bristles. We simply walk down the rows of potatoes, brushing the plants with downward strokes, causing the larvae and

More recently, we've tried a new, low-tech approach: we confront our enemies by hand.

sometimes their parent beetles to roll into the tubs. In this manner, we catch them by the thousands and dispose of them in ways I will not discuss here. Those that miss the tubs and roll onto the soil, because they are slow crawlers and not well endowed mentally, have a hard time finding their way back to the dinner table. I'll grant it's a primitive approach to pest control, but so far, it has achieved the desired result. Five or six of us working the broom heads for a couple of hours can make a major difference.

By mid-July we're ready to start digging and selling (and eating) the fruits of our labors. For most of us this is a long-awaited day. Nothing can quite compare with the first new potatoes of the season. We dig them in pairs, with a combination of forks and fingers—one person using the fork, the other his or her fingers to riffle through the soil and extract the bounty within. We start with the Red Golds since they are an early-season variety.

During the first couple of months of harvesting, we try to gauge market demand on a day-by-day basis and dig accordingly. In the heat of summer, spuds store well in the cool soil and it's nice to be able to promote them as freshly dug. But if they stay in the ground too long they may suffer the nibbling of voles or the burrowing of wireworms, neither of which enhances their retail appeal. Also, if left too long in wet, warm soil, potatoes may even begin another generation of growth.

In late September, well before the ground freezes, when the soil has a little moisture in it, but not too much, we undertake a major potato harvest. For this we mobilize all hands and make use of a tractor and an implement called a middle buster. The tractor pulls the middle buster, which undercuts the potatoes, one row at a time. By now the plants' foliage has browned and died back, and if there are weeds present, it can be difficult from the seat of a tractor to know exactly where the buried tubers are. When this is the case, someone is assigned to walk in front of the tractor to direct the driver and check the ground periodically with a fork. If all goes well, most of the spuds are kicked up to the surface, undamaged, but often quite a few remain under the soil and must be uncovered

It's a Long Road
to a Tomato
130

by human hands or brought up with forks. It can take ten or twelve hours, usually spread over a few mornings or afternoons, before we have all our potatoes safely crated and stored in the barn's root cellar. Once this job is done, I'm able to get a fair estimate of the actual yield for the season.

If stored in a cool, dark place with about 50 percent humidity, healthy potatoes will retain good eating quality for at least nine months. They are a crop that a farmer can take to market right through the winter.

And so, after my initial snobbish reluctance, I have become a grower of potatoes—and am glad to be one. They please me in a special way. Most of our crops grow on the surface for everyone to see. A few, including turnips, onions, and radishes, reside partly in the soil and partly out of it. Carrots are primarily earth dwellers, but not entirely—they always leave a little of their orange crown exposed so you know where they are. Garlic, of course, grows proudly underground, but it sports a sturdy stem above the surface.

Of all our crops, only potatoes seem true denizens of the soil. There is an earthy substantiality about them. Digging down into rich, moist soil to uncover a vibrant cluster of potatoes, some still connected to the mother root, is a singular experience. Perhaps it is true that potatoes are a food of the masses, but I have come to understand that it is the wisdom of the masses to eat them.

Perhaps it is true that potatoes are a food of the masses, but I have come to understand that it is the wisdom of the masses to eat them.

KURI ENCOUNTERS a PORCUPINE

I *have never* seen a porcupine in Greenville, nor met anyone who has. I'm told we're at the southern tip of the porkie's range and that the prickly creatures are seldom found in these parts, though they're common in the Catskill Mountains to the north. But now and then a wandering soul, in search of a mate or a winter den, perhaps following the line of the Shawangunk Ridge, must arrive in our small, rural town. And, of course, it stands to reason that my dog, Kuri, with his ever-curious nose and passion for the hunt, would eventually come across such a one.

It was midmorning on a summer's day. Kuri, who would normally accompany us into the fields at the beginning of the workday, had been absent for at least a couple of hours. I didn't think much of it, and, at any rate, was focused on the challenging task at hand—namely, trellising peas that had been left too long and were already beginning to lie down on their sides. While engaged with a particularly difficult row of sugar snap peas, I heard some wild fits of barking coming from the woods on the ridge. Probably he's got a raccoon up a tree, I thought.

A half hour or so later, Kuri showed up on the lawn and I noticed with a sideways glance that he was acting a little strangely. He seemed to be brushing his face with his front paws in a highly distracted manner. I walked over to take a closer look and saw at once the reason for his unusual behavior. His face had become a veritable pin cushion, with a dozen or more whitish, black-tipped quills sticking out in all directions. I immediately commiserated with him and understood from his body language and strained facial expressions that he would be open to receiving any assistance I might be able to offer.

With my left hand, I took hold of him by the collar and with my right I tried to extract one of the larger quills from his nose, but without success. The sides of the quill were smooth and difficult to grip and, the more I pulled, the less willing Kuri was to accept my help. In one violent attempt to free his head from my hold, the quill snapped. Most of it was left in my hand, but a small amount remained deeply embedded in his unfortunate nose.

At this point I understood that his predicament would not be so easily resolved. My choice of the word *predicament*, while it may seem unusual, is deliberate. Being a longtime fan of spoonerisms (the usually accidental switching of syllables between words, as in "a crushing blow" to "a blushing crow"), I've waited for years for the right occasion to use this interesting word for what I perceive to be its spooneristic value. By rearranging the order of the first two syllables, one arrives at the entirely new, but highly appropriate, word *depricament*. Short of coming upon a case of male genital mutilation or dismemberment, which I have no desire to do, this may be the best opportunity to use it that I'll ever have.

I encouraged Kuri to relax by rubbing his shoulders and back. I then guided him over to his doghouse, where there is a chain that is used to constrain him when necessary. He sat down and allowed me to attach the chain to his collar. At that moment he looked rather contented, feeling, no doubt, that anything was preferable to the kind of help he had just experienced.

I ran off to fetch a pair of blunt-nosed pliers and on the way decided to grab a camera as well. By now my wife and a seasonal farm intern had appeared on the scene. Both were horrified at Kuri's condition and urged that he be taken to a vet immediately. A short argument ensued, but I prevailed, confident that a suitable "depricament" could be undertaken with their help. Of course I was made to feel some guilt, but counted that a small price to pay for what I was about to do for my dog.

First, I asked Flavia to take a few photographs, because I felt sure that never again would Kuri's face be so adorned and that a record should be had for posterity. While it may seem callous to think of taking photographs during such an emergency, I would argue that Kuri has led a highly eventful and action-packed life. He has pursued adventures and misadventures with equal enthusiasm and is no stranger to injury and the resulting pain and discomfort. Kuri has always sought out heightened canine states, especially those related to pursuit of his quarry, and he appears to relish them all, regardless of outcome. I am inclined to believe that he takes a certain pride in his battle wounds and would fully approve of my interest in recording them, when feasible.

Ready to proceed, I put on a pair of leather gloves to protect myself against fast-moving teeth. I then sat down beside Kuri and began rubbing him affectionately. Before long he rolled over onto his back and willingly accepted my attention, as he has done so many times before. I stroked his tight, lean belly and talked to him in soothing tones until it seemed that he had forgotten his misfortune and was happily immersed in the simple pleasures of a summer's day.

I then wrapped my left arm around him in a neck hold and held him firmly to my chest, while my wife and the intern each took hold of one of his back legs. Constrained in this manner, Kuri had no choice but to submit to my plan. Carefully positioning the pliers in my gloved hand, I went for my first quill, the one that had already broken. I closed down on it about an eighth of an inch from the flesh of his nose and pulled up as straight as I could. Kuri's neck muscles tightened as he tried to free himself and his front legs paddled furiously in the air, but to no avail. The three-person grip was more than he could overcome. The quill put up some measure of resistance, then, almost unexpectedly and to my great relief, popped out. It dawned on me then: "These damn things are barbed! But no matter, they come out anyway."

I put on a pair of leather gloves to protect myself against fast-moving teeth.

Later, upon closer inspection of one of the larger quills, I could not see any barbs, and when I ran a finger along the quill, from the tip to the stem, it felt smooth to the touch. But, when I ran my finger the other way, from the stem to the tip, I detected a definite roughness and resistance. There were barbs all right, but they were microscopic and all the more diabolical for it. I have since learned that porcupines have few natural enemies. Most predators know better than to attack these lumbering, slow-moving creatures. While easy enough to run down, they are quite capable, via a lightning-fast flick of the tail, of delivering any aggressor a prickly and sometimes deadly rebuff.

But back to Kuri and his ongoing "depricament." After the first quill was removed I relaxed my grip on him and rubbed his belly some more, telling him all the while what a fine and brave dog he was. Then I took careful aim at the next quill, one that had penetrated his lip. Again, his neck muscles tightened and the quill resisted and then came out. The sense of release was profoundly satisfying. Within five minutes, all the quills were out and a couple of them set aside as souvenirs. Unchained, Kuri took a long drink of water, then sat down on the lawn under the shade of the mulberry tree as though nothing disagreeable had happened. All that was left to tell of his ordeal were a few pinpricks of blood on the tip of his nose.

To my knowledge, Kuri has not encountered another porcupine. Or, if he has, he's allowed it to pass unmolested. I'd like to believe that his intelligence, his memory, and his self-restraint are sufficiently developed to make this possible. While I do have some confidence in his mental capacity and his memory, I may be crediting him with more self-control than he possesses.

A DAY at the MARKET

3:10 AM ☙ The alarm goes off. It's Saturday morning—market day. Four and a half hours sleep is all I'm going to get. I stop the alarm and get up reluctantly, but without delay. To linger would be dangerous. After a bowl of cereal, a cup of green tea, the requisite ablutions and mandatory stretching exercises, it's time to board the Mitsubishi and head for New York City. Don't forget the lunch box and plenty of fluid. A long day lies ahead. And probably a hot one.

Two of my helpers for the day are waiting by the truck. We exchange brief greetings and take our seats. As we pull out of the driveway, the sky is full of stars and there's a new moon to the east cradling the old moon in its arms. At first light of dawn a hazy stillness lies over the land. There's hardly a vehicle on the road as we drive by fields of freshly mowed hay and shoulder-high corn. Shapes of cows loom on the crest of a dark hill. A red fox with an impressive bushy tail and determined gait crosses the road in front of us. For a while we ride in silence, immersed in our own thoughts, perhaps revisiting dreams that were in progress when our alarm clocks called us to work.

I fight off drowsiness with a shot of black coffee and try to engage one of my crew in conversation. Staying alert at the wheel in the middle section of the trip can be a challenge, especially when your passengers start nodding off beside you.

We proceed through sleepy towns and villages, past the lakes and reservoirs of northern New Jersey, then up the big hill known as Skyline Drive, where we catch our first glimpse of New York City. The transition is under way—from farm and field where nature is paramount to the incredible human bustle and activity of Union Square, Manhattan. I can feel the change taking place in my body: a slight tensing and tingling in my stomach, a mild sense of anticipation and excitement. The day that lies ahead will tax to the full our bodies, our brains, and our social skills. From a bare piece of asphalt in the heart of the world's greatest city, we will create and adorn our own retail space and engage in commerce and conversation with an almost endless stream of people.

6:00 AM 🐦 We enter the Lincoln Tunnel; fifteen minutes later we take a left off Broadway onto 17th Street and pull into Union Square. A few dozen farmers have already arrived and are busy unloading trucks, putting up canopies and signs, setting out crates filled with fruits and vegetables, carrying buckets bursting with flowers and coolers packed with meats, cheeses, and fish.

Situated at the north end of Union Square Park, between an orchardist and a flower grower, our space has a frontage of twenty-four feet and a depth of eighteen feet, plus room behind for our truck. We'll use every square foot allotted to us and would be happy for more, but so would our neighbors. There's a premium on real estate in Manhattan, especially at Union Square Greenmarket on a Saturday.

Once out of the truck, we go straight to work. First we must build shade and shelter. Our canopy setup requires thirty-six poles, sixteen corner and connecting pieces, one silver tarp, two mesh shade tarps, an engineering degree, and a fair amount of heavy lifting. Once the canopy is in place and secured with ropes and sandbags, we set about creating a display of our vegetables and herbs. We use tables, crates, boxes, tubs, and numerous signs; we design walkways for easy access and visibility; we hang plastic bags of varying sizes from canopy poles; we set up scales and a cash box and we strap on money belts for roving sales. We work to create a customer-friendly environment that will draw people in, encourage them to browse, make their selections, and then come to the main table where items are weighed and bagged and cash changes hands.

7:15 AM ❧ The early birds start arriving. People are smiling and happy to be out while there's a little coolness in the air. We're not quite ready for them, but we're moving fast. Between topping off display crates, posting signs, and arranging herbs, we manage the first quick sales of the day.

We have a good load for late August: potatoes, red and yellow onions, squash, shallots, garlic, peppers, beans, fifty-odd crates of lettuces and assorted other greens, masses of basil and parsley, and at least a dozen other herbs. Best of all, we have ripe tomatoes—a good six hundred pounds of them, of varying shapes, sizes, and colors, all picked on Friday and rich with the flavor of sun and earth. We know they taste good. We've already eaten plenty of them.

8:00 AM ❧ The pace is quickening and the temperature is rising. A half-dozen people are at the stand, plucking lettuces from display crates, filling bags with potatoes, garlic, heirloom tomatoes, and sweet salad onions. Most of our customers have been coming to the stand for years. We know their faces well and often their

I can feel the change taking place in my body: a slight tensing and tingling in my stomach, a mild sense of anticipation and excitement.

names. Greetings are exchanged, handshakes, smiles. Roger Boswarva, my Australian mate, and his girlfriend, Virginia Koenig, show up with their usual infectious good humor. Virginia gives me a hug and Roger gives me the tape of a rugby game between Australia and New Zealand, in which the latter, my home country, presumably loses. They stock up on vegetables for the week and head off for bread and cheese at other stands.

We work as a team. There are four of us now. (Our fourth has just arrived—because there's room for only three in the truck, one person has to take the bus down to Port Authority, then catch a subway over to the market.) Two people work behind the main table, weighing, bagging, doing lots of arithmetic, and handling most of the cash. One keeps the smaller herb table looking good and replenishes the display crates on the right side of the stand. To preserve freshness and aroma, we keep most of our herbs in coolers and leave only a few of each on display at any given time. This means that the herb person must constantly restock our rustic wooden displays.

The fourth person handles the left side of the stand and takes charge of the tomatoes—no small task. The different varieties are set out on several large bread trays positioned in the front and middle of the stand. There are large reds, small reds, plums, yellow low-acid tomatoes, and three different types of heirlooms. Orange and yellow cherry tomatoes are sold in pints at the main display table. All of them need constant monitoring and replenishment. The key to selling lots of tomatoes is bountiful displays of good-looking, good-tasting fruit. We lose a few to indecisive shoppers with overactive thumbs. These are quickly culled out and sold later in the day as seconds.

11:00 AM ☙ It's close to ninety degrees out, but the activity at the stand is intense. We haven't stopped moving for five hours. My stomach is rumbling and wondering why, in the midst of so much food, it is being deprived. It'll have to wait. There's a line of at least ten people with produce in their hands. They haven't come to the market to watch me eat. But they seem happy enough to be waiting

The fruits of our labors and our fields are now dispersed across New York City.

It's a Long Road
to a Tomato
140

under the shade of our canopy, chatting among themselves and with us. It's rare that anyone shows impatience. If someone does, other customers soon cast disapproving glances in their direction.

Around noon ❧ Colin Alevras, chef/owner of the Tasting Room on First Avenue and 1st Street, visits our stand and picks up an assortment of herbs and a few pounds of shallots. He asks how my persimmon tree is doing and places an advance order for one perfect persimmon on a branch about six inches long. Then, Peter Hoffman of Savoy stops by with his trademark bicycle/shopping cart, laden with fresh produce—presumably for his restaurant's evening menu. We talk politics for a few minutes, and then along comes Sally Schneider, another regular and author of *A New Way to Cook* (a copy of which is always on hand at our stand). Sally and Peter know each other and immediately strike up a conversation. I duck away and tend to the basil display, which I notice is empty, and fill a few pint containers with green beans. When I get back to them, Peter is holding a bag of Cherokee Purple tomatoes and several bulbs of Rocambole garlic; Sally has her usual two pounds of mesclun and several other items.

1:30 PM ❧ It's got to be over ninety degrees and it's humid. We're beginning to show a little wear and tear, and so are our customers. There are fewer of them, and their smiles are less convincing. We take turns resting and eating. I sit on the back of the truck for ten minutes and consume a tuna fish sandwich with a handful of mesclun and a hard-boiled egg. This gets washed down with copious amounts of cold water spiked with orange juice and lemon. My stomach approves.

Between 2:00 and 3:00 there's a definite lull in business. It's too damned hot. We spruce up the stand a bit and start loading empty crates and coolers into the truck. We rearrange some of the displays. The lettuces that haven't sold yet are under assault from the sun, which is now hitting us frontally and taking a toll on anything fresh and green that is not protected by the canopy or shade tarps. I'm feeling weary and would love nothing better than to lie down in a cool room. But that's not going to happen. Instead, I wander off for a while and talk about tractors with a friend and neighboring farmer, Andy Van Glad, who sells maple syrup.

4:00 PM ❧ Activity at the stand picks up. The late-day crowd is moving through the market. We still have a couple of hundred pounds of tomatoes, but they're selling briskly, as are the garlic and onions. The remaining greens are not looking good, and we start cutting deals. At five o'clock we put the mesclun on sale; the last two coolers move quickly to end-of-the-day bargain hunters.

6:00 PM ❧ Time to break down. We're all tired from the long day, the hard work, and the heat, but the prospect of sitting in an air-conditioned truck and driving home spurs us on to this final task. It takes over an hour to dismantle and load everything up. We keep getting interrupted by passersby wanting to make last-minute purchases. Where possible, we oblige. City Harvest shows up with large plastic bags. We give them thirty or forty pounds of molested and very ripe tomatoes, a dozen heads of lettuce, and some other greens, and are reminded that not everyone in the city has food to eat.

On the way home we are in good spirits. We feel lighter and pleasantly gratified. The fruits of our labors and our fields are now dispersed across New York City. We imagine people sitting down to dinner, remarking on the freshness and flavor of what they are eating, glad in the knowledge that there are still farmers' markets in their city—and still local farmers to provide them with decent food.

9:30 PM ❧ We pull into the driveway. The sky is full of stars again. I feel a deep sense of relief. A startled rabbit runs for cover. As we make our way down to the house and barn, the truck headlights catch the eyes of Aldo and Tiki, our big white dogs, as they come running from their sentry positions to greet us.

It's a Long Road
to a Tomato
142

Aldo and Tiki—the Sentinels

BRAVE NEW VEGETABLES

I *saw a* cartoon the other day. The caption read: "Two out of three farmers favor genetic engineering." Above the caption was a cartoon representation of one farmer who was opposed to genetic engineering and another who was in favor. Each was holding a pitchfork, but the farmer in favor of genetic engineering had two heads. It gave me a good laugh. But it was the kind of nervous laugh that comes from finding something both funny and disturbing at the same time.

As a small farmer committed to a sustainable and land-friendly form of agriculture, I'm uneasy with what I hear and what I read. The genetic manipulators of plants and other organisms are seeking a level of control and ownership of our food production that is without precedent.

The brave new world that we have so rapidly entered is a world in which the genetic codes and building blocks of life—the very stuff of evolution—are being altered, manipulated, and controlled by scientists. Invariably, these travelers into the future are well-paid workers in white lab coats, in the employ of large transnational corporations. What they will come up with is almost beyond imagining, but we're beginning to get a few early glimpses.

Monsanto's New Leaf Potato, introduced in 1995 and withdrawn from the market in 2001, is a good example. Here was a superspud with a bacterial pesticide, in the form of Bt, spliced right into the plant's DNA. Unsuspecting Colorado potato beetle larvae, long the bane of potato growers, would take only a few bites out of this plant before they curled up and went to sleep for good. Not such a bad thing, you might think, so long as you didn't mind ingesting the *Bacillus thuringiensis* bacteria in every one of these spuds you ever ate. Once the genetic manipulation occurred, the Bt genes were produced by the plant and ended up in every cell—in the tubers, the leaves, even the pollen.

Bt was not new, nor was it discovered by Monsanto. For many years, organic growers have used a natural form of this bacterium as a biological (not chemical) pesticide. Its use is permitted under organic standards (some Bt is artificially produced and is not permissible—there are clear rules about which Bt can be used under organic regulations and which cannot). Bt breaks down rapidly in the soil and affects only target pests. It is sprayed on potato plants when Colorado potato beetle infestations are too high. For me, that used to be once or twice a year, or not at all, if I didn't see many of the beetles' larvae. Lately, as I've already mentioned, we've stopped spraying Bt altogether in favor of those carefully timed removals of the pest by hand.

But for those growers who choose to spray Bt, the future did not look good once Monsanto's New Leaf Potato became commercially available. Because so many beetles were consuming Bt via this

Radishes

genetically altered spud, it was widely believed (and even conceded by Monsanto) that the bugs would eventually develop resistance to the pesticide. Sooner or later a Colorado potato beetle would come along for whom Bt was just a tasty condiment. This individual would one day find a similarly endowed mate, and the two would have offspring who would likely share the same resistance to Bt as their parents. The offspring would, presumably, proliferate. If this had happened, the Bt used judiciously and only when needed by organic growers would no longer work, and we would have lost one of the most benign and useful pesticides we have.

Poor sales, high expense, and rejection by McDonald's as a source for their french fries prompted Monsanto to withdraw its New Leaf Potato from the market in 2001. That was a relief. But the technology still exists, and it may come back to bite us when market conditions are more favorable. In the meantime, both cotton and corn with Bt spliced into them are widely in use.

A different aspect of the genetic rearrangement of our food, but one that is perhaps more scary, especially for farmers, has been labeled the "Terminator" technology. This was developed in the late 1990s by Delta and Pine Land, a Mississippi biotech company working with USDA researchers and aided by a government grant. The "Terminator" allows a plant's genetic makeup to be altered in such a way as to ensure future sterility. This means that if you're a farmer (or gardener) who likes to save seed and plant it the following year, you won't have any luck. The "Terminator" eliminates future generations. You get just one season. After that you go back to the seed company.

In March 1998 a patent was granted for the "Terminator" technology, and soon afterward Monsanto attempted to acquire Delta and Pine Land Company with the presumed intention of employing the Terminator to prevent "unauthorized use" (i.e., saving and replanting) of its proprietary seed. Public outcry and Monsanto's fear of further damage to an already tarnished reputation led the company to withdraw its offer and publicly state that it had no further interest in using the Terminator.

But, like the New Leaf Potato, the "Terminator" has not gone away. In fact, it has been further refined and developed and, in the process, lumped in with several other genetic-modification techniques. Genetic Use Restriction Technologies, or GURTs, is the term that is now being used to describe the use of chemical inducers to control the expression of various genetic characteristics in plants. Plant color, shape, and size, and even tolerance to cold, heat, and drought are all traits that may be influenced. So is sterility in the next generation—the old Terminator in a new guise and with a less-ominous-sounding name.

Monsanto holds patents for GURTs but is not the only company that is showing an interest in these new technologies and their market potential. Patents for GURTs are also held by Delta and Pine Land, DuPont, Syngenta, as well as the USDA and several American universities. The range of possible applications is broad since the GURTs patents usually apply to both transgenic and conventionally bred seed. Crops such as wheat, rice, soybeans, oats, and sorghum are all likely candidates for GURTs.

By patenting newly developed plants and controlling plant reproduction and other characteristics, seed companies hope to expand their competitive edge in the global seed market. This could be lucrative in North America and other parts of the Western world, and even more profitable in less developed countries where farmers have relied heavily on saved seed to go from one year's crop to the next. The impact, especially of induced sterility, on farmers in developing countries could be devastating. As the new hyper-seeds are vigorously promoted, local varieties are likely to disappear and small farmers will have little alternative but to return to the seed company year after year. If they can't afford the patented, transnational seed prices—a distinct possibility—they may be forced to sell or abandon their land, which will result in further consolidation of agricultural production, and, most likely, further population influx into already overcrowded third-world cities.

The gene manipulators claim that their new plants will offer farmers many advantages. There may be some short-term advantages, but I am skeptical about real long-term benefits. And there is the looming question of what unknown effects the genetic modification of our food might have, what surprises might lie in store for us and for the planet. The emergence of highly resistant, super-competitive insects and weeds is a possible outcome that could make farming without chemicals a lot more difficult. Other concerns voiced by scientists include: the development of new toxins and allergens in our food; the unintended transmittal of genetically modified characteristics to other, related species; the movement of diseases from one species to another; more intensive use of agricultural chemicals; loss of biodiversity; and general ecological disturbance. Each of these is reason enough for caution.

Being a strong believer in diversity and a light touch, I would rather work with nature than try to control her at every step. I have found this approach to yield very satisfactory results, despite the oft-heard claim by the purveyors of GMO seeds and agricultural chemicals that "organic farming is an elitist, fringe movement that could never feed the world"—a claim I would strenuously dispute. And recent findings bear me out: David Pimentel, a Cornell professor of ecology and agriculture, with collaboration from the Rodale Institute and the USDA, compared the yields of organically and conventionally grown corn and soybeans over a twenty-two-year period. He and his colleagues found almost no difference in yields between the organic and conventional methods, but noted significant soil degradation and erosion on the conventional plots (see *Bioscience*, volume 55:7, July 2005).

More recently, a major study conducted by the United Nations Environmental Program and the UN Conference on Trade and Development concluded that small-scale organic agriculture is more conducive to food security in Africa than most conventional production systems, and is more likely to be sustainable in the future. The study's report, titled "Organic Agriculture and Food Security in Africa," was published in 2008.

Genetically engineered plants are, I fear, just another step down the same conventional corporate road. They promote monoculture and erode diversity. They will further concentrate control of our food supply in the hands of a few large companies. Farmers, the traditional custodians of the land, will make fewer and fewer choices about what they grow and how they grow it.

Already there are genetically altered plants on the market that have herbicide resistance built into them. Monsanto's Roundup Ready soybeans are capable of withstanding a good drenching with the ubiquitous herbicide Roundup, also a Monsanto product and one that accounts for sales of over $2 billion annually. Almost all plants and weeds die when sprayed with this herbicide (and, lately, it would seem, frogs and tadpoles as well), but not Roundup Ready soybeans. In this case the farmer becomes dependent on the seed company, not only for the seed, but also for the herbicide that goes with it. Talk about getting hooked. The development of plants that can withstand pesticides leads to more pesticide use, not less.

The genetic manipulation of our food is taking place with minimal regulatory oversight and virtually no input from the public. (This is not the case in Western Europe and many other parts of the world, where the biotech industry has encountered strong opposition to its new "frankenplants.") You might say that we in

The companies who are bringing us these genetically modified foods are the same ones that have been peddling pesticides for the past fifty years.

the United States have been kept very well uninformed. Most people's knowledge of the subject is scant indeed. Monsanto has fought fiercely and successfully against any attempt to label its new transgenic products, and the federal government has not seen fit to challenge them. Certainly, there has been a dearth of regulatory scrutiny. Our ability to vote with our wallets, to buy or not to buy, based on full disclosure of relevant information and our own individual preferences, has been seriously eroded.

I suppose it's easy to be alarmist about these radical developments in our food system. Perhaps they will bring us into a brighter future. Perhaps I'm just a skeptic and a Luddite. But then I recall that the companies who are bringing us these genetically modified foods are the same ones that have been peddling pesticides for the past fifty years and telling us how lucky we are, while neglecting to mention that many of their past products are now banned due to harmful effects on our health and environment that became evident only with time.

To be honest, I don't expect to see two-headed farmers anytime soon, but nor do I think it wise to put blind trust in an unrestrained, profit-driven, corporate culture—especially with something as vital to our well-being as the food we eat.

The development of plants that can withstand pesticides leads to more pesticide use, not less.

It's a Long Road
to a Tomato
150

PUTTING IT BACK

n late August, we sowed a cover crop of clover where the garlic

had been. It will prevent erosion in the winter and put nitrogen

into the soil for next year. Aided by warm September days and an

inch or more of rain a week, it's growing nicely. From the upstairs

bedroom window, it glistens a fresh young green in the late-day sun.

At night, if you shine a powerful flashlight on the field, sometimes

you can see the luminous eyes of deer.

As an organic grower for whom synthetic chemicals are not an

option, I'm sometimes asked by my Greenmarket customers what I

do to maintain fertility. It's a fair question and one that deserves a thoughtful answer, but usually, in the flurry of selling vegetables, time does not permit. I rattle off a couple of lines about compost and rock powders, then return to the business at hand, namely, extracting the questioner's money as efficiently as possible in exchange for my produce. This gives me a certain satisfaction, but I'm often left feeling that I could have said more—that, in a sense, I slightly shortchanged my customer.

Regardless of whether a farmer (or gardener) grows organically or with the aid of chemicals, the question of maintaining fertility has to be addressed. Plants have four major requirements: sunlight, air, water, and nutrients. While the amounts of sunlight, air, and water (supplemented by irrigation) are more or less stable from one season to the next, the soil's supply of plant nutrients can be depleted after just a few years of continuous cropping. Every time a vegetable—say, a head of broccoli—is harvested and removed from the farm, some nutrients and minerals go with it. We assimilate these into our bodies when we eat the broccoli, and benefit accordingly. But at some point, what was taken from the soil must be given back.

The conventional farmer usually makes up the difference with synthetic petrochemical fertilizer, the principal ingredients of which are nitrogen, phosphorus, and potassium. The amounts of fertilizer used by farmers and the proportions of each ingredient are often based on the results of soil tests and what an individual farmer believes is needed for a particular crop. There is a definite tendency to overfertilize, just to be on the "safe side."

Fertilizer ratios are always expressed in the same order: nitrogen (N) first, then phosphorus (P), then potassium (K). A bag of conventional fertilizer, labeled 20:10:20, will contain 20 parts nitrogen, 10 parts phosphorus, and 20 parts potassium, all in a highly water-soluble form, readily available to plants. A bag of NPK 10:20:10 will have the proportions reversed (twice as much phosphorus and half the amount of nitrogen and potassium). A lot of fossil-fuel energy is needed to create the chemical reactions necessary to produce this kind of fertilizer.

Other important plant nutrients are calcium, magnesium, sulfur, and carbon, the last of these coming primarily from atmospheric carbon dioxide. Plants also need trace elements or micronutrients such as iron, manganese, boron, copper, zinc, and molybdenum, but these are required in very small amounts and are usually present in adequate quantities in most soils.

Organic farmers rely on the same list of nutrients but prefer to make them available to plants through natural processes, over a longer period of time. "Feed the soil, not the plant" is a favorite maxim of organic growers. This is best achieved by adding organic matter to the soil, usually in the form of compost or animal manures and sometimes by supplying rock powders that contain essential minerals. The idea is to encourage microbial and biological activity.

Under our feet there is a vast subterranean community—huge populations of different organisms engaged in a complex ecology that rivals the plant/animal relationships found on the surface of the earth. A handful of healthy soil can contain more than a billion organisms—among them bacteria, fungi, protozoa, nematodes, and, of course, earthworms. The combined weight of these organisms in the top six inches of an acre of soil can be as high as ten tons.

Most soil organisms make their living breaking down organic matter and converting it into a form that is usable by plants—they are in the recycling business. Others are predatory, feeding on the primary decomposers of organic matter. All leave behind humus, a sticky, dark brown substance that holds together the mineral particles in the soil, without which soils are highly vulnerable to erosion. Humus also holds water and air and generally makes for a light, fluffy soil that plants love.

Compost can be made from most plant-based, carbonaceous matter. Materials rich in nitrogen, such as animal manures, are also useful. For backyard gardeners, grass clippings, leaves, green kitchen scraps, and the like will work quite nicely. On the larger scale of a farm, a great deal more raw material is needed. The bulk of the ingredients in my compost are obtained from two neighboring farms—one with horses, the other with cows. The horse farm uses

Every time a vegetable—say, a head of broccoli—is harvested and removed from the farm, some nutrients and minerals go with it.

wood shavings as bedding in the horse stalls. When workers at the farm clean out the stalls, the shavings are peppered with horse turds and soaked with horse urine. The turds are a great natural source of nitrogen, phosphorus, and potassium, as well as other nutrients, and the urine is rich in nitrogen, albeit in a highly volatile form.

The use of horse manure to grow vegetables has a certain nostalgic ring for me. It takes me back to my distant youth in New Zealand. When I was a young boy growing up in a suburb of Wellington in the late 1940s, the milkman would come by each morning with a horse-drawn cart and crates of bottled milk. The horse was trained to stop and wait outside each house while the milkman gathered the empty bottles people had put out the night before. Depending on how many milk tokens were left for him, he replaced the empty bottles with full ones.

It was a good system: no noisy, polluting engine, no need to constantly get in and out of a vehicle—just a signal from the milkman to the horse and the clip-clop of hooves.

On occasion, as one might expect, the horse would leave behind a pile of fresh manure on the street near our house. My mother, an avid gardener, was always on the lookout for such an event and would unabashedly rush out with a shovel and bucket to bring back the golden nuggets for her compost pile. Unfortunately, our next-door neighbor, Mr. Wilson, was also a gardener and he also coveted the horse's droppings. He was an older gentleman who lived with his very ancient mother. As a young man, while working for the New Zealand Railways, he'd suffered a terrible accident which left him with two wooden legs. We could hear him cautiously making his way down the stone path from his house, sounding almost like the clip-clop of the horse, but slower.

The competition between Mr. Wilson and my mother for the horse's droppings was a standing joke in our house. One of my older sisters claimed that when our mother heard the heavy wooden footsteps on the path in the morning, around the time of milk delivery, she would look out the front window and, if the horse had obliged, leave by a side door and claim the prize before

Mr. Wilson got there. (I strongly doubt that my mother ever actually did this, but the story, nonetheless, illustrates the lengths to which she might have gone for the sake of her compost.)

But that was long ago, and it was pure horse manure. What I receive today is a lot less concentrated—in fact is mostly softwood shavings, which are low in nitrogen and very high in carbon. Good compost should have a carbon to nitrogen ratio (C:N ratio) of around 20 or 30 parts carbon to 1 part nitrogen. What the horse farm gives me is definitely higher in carbon than is desirable, and that's where the cow manure comes in. It provides the extra nitrogen my crops need.

When everything is going according to plan, I am able to coordinate deliveries between the horse farmer and the dairy farmer. Ideally, we lay down several loads of horse bedding to create a pile or windrow about eighty feet long, ten or twelve feet wide, and two feet high. On top of this is spread a load of cow manure a few inches thick. Then comes another load or two of horse bedding, then another layer of cow manure, and so on.

When the pile is four or five feet high, I start turning it with the front-loader bucket on my tractor. This is done about once a week. It brings air and moisture into the center of the pile, mixes the horse bedding and the cow manure, and generally speeds up the composting process. I like to do this work in the summer months when the decomposing and recycling organisms are most active. The goal is to have the pile heat up to temperatures above 130 degrees, which will kill most of the weed seed and pathogens that might be in the manure. Each time the pile is turned, steam rises out of it due to the heat that is released from the inside. The compost is finished when it has a crumbly texture, a dark, pleasing color, and almost no odor.

For me, the process can take nine months or more, depending on the proportions of the mix and how thoroughly it is turned, and because wood shavings take a relatively long time to break down. (Compost made with greener, softer plant material can be ready much sooner.) When the time is right, I spread the finished compost with an old manure spreader, then incorporate it into the soil with a chisel plow.

I make between fifty and one hundred tons of compost each year and apply it to most of my productive soils at a rate of twenty to thirty tons per acre every two or three years. Each year I spread compost in different fields, giving more to those fields that are cropped more heavily and more frequently.

Compost is the foundation of my soil-building and fertility program, but compost alone does not solve every problem. Due to the vagaries of glacial movement long ago and a host of other geologic and topographic factors, there are several distinct soil types on our farm.

The *upland* or *mineral* soils are derived from the gradual weathering of rocks over long periods of time. These soils are naturally high in minerals and low in organic matter. They benefit greatly from the addition of compost.

Muck soils or *black dirt,* the other soils on the farm, were formed over millennia in shallow lakes, marshes, and depressions. These soils, which derive largely from decomposed plant and animal remains, are naturally high in organic matter and nitrogen, but low in minerals. They have less need for compost but benefit from the periodic addition of rock powders, specifically rock phosphate (a natural source of phosphorus) and greensand (a source of potassium).

When it comes to pH and soil chemistry, the picture gets a little murkier. The two or three soil types on my farm that are most suitable for vegetable production require periodic applications of calcium carbonate or limestone. The limestone serves two purposes: it adds calcium, and it keeps the pH, or acidity, of the soil within a range (about 6.3 to 6.8) that maintains the right nutrient balance for plant growth. When the pH is too low or too high, certain nutrients attach chemically to other elements in the soil and become unavailable to plants. For example, even though there may be plenty of phosphorus in the soil, if the pH is too low (i.e., the soil too acidic), the phosphorus will bond with iron and aluminum and become immobilized.

In addition to the compost, limestone, and rock powders, I sometimes use small quantities of purchased organic fertilizer, largely as insurance. These fertilizers contain varying amounts of nitrogen, phosphorus, and potassium, just like their chemical counterparts, but the ingredients come in natural forms. Because natural fertilizers are less water soluble, they are not as rapidly available to plants. Nor are they harmful to soil organisms, as synthetic fertilizers can be. And they are much less likely to leach into groundwater and streams, which is another problem with the synthetics. The natural fertilizers are largely assimilated into the soil before being ingested by plants. Their value is more gradual and benign.

There are still other tools available to organic growers in search of fertility. Fish emulsion (which is rich in nitrogen) and kelp (a natural source of micronutrients) are sometimes used to enhance plant growth. They are usually applied as a foliar feed, sprayed directly onto plant leaves. Unlike most other items in the organic grower's fertility chest, they can be assimilated by plants quite rapidly. We use them primarily in the greenhouse to keep our seedlings healthy, before they are transplanted into the field.

Finally, cover crops and green manures should be an integral part of any organic farm's soil-building and fertility program. There are plenty to choose from; we use several different ones each year. Buckwheat, for example, grows rapidly in the summer heat—it is good on fallow ground, to keep the weeds down. Winter rye and oats provide ground cover in the winter and prevent erosion. Hairy vetch and assorted clovers are good for long-term soil improvement and are able to capture atmospheric nitrogen and convert it into a form usable by other plants. All of these cover crops contribute organic matter to the soil and confer numerous other benefits.

Most of us who work closely with the soil become aware over time of what a dynamic and life-sustaining resource it is. To nurture it as it nurtures us and to use it well become natural priorities. To be on good terms with the soil is to be on good terms with the earth. In one form or another, willingly or not, we all find our way back to it.

To be on good terms with the soil is to be on good terms with the earth. In one form or another, willingly or not, we all find our way back to it.

THE DRIVEWAY RABBITS

What is man without the beasts? If all the beasts were gone, man

would die from a great loneliness of spirit.

— Words once attributed to Chief Seattle
of the Suquamish Indians (but now
thought to be fictional)

On our first visit to the farm, in the company of a real estate broker, we saw a rabbit on the driveway. It ran ahead of the car ten or twenty feet in a zigzag fashion, as rabbits often do, then stopped abruptly in our path, apparently unable to decide which way to proceed. A moment later, its mind made up, it hopped through an opening in the high grass along the driveway edge.

I took it to be a good sign, this rabbit sighting, though I can't exactly say why. Perhaps it took me back to my boyhood days and the fascination I had then with small creatures that run off into their own secret, hidden worlds (a fascination I still have vestiges of today).

Passing the spot where the rabbit had disappeared, we continued down the long, dirt driveway into the heart of the farm. Over the next few hours we eagerly explored the possibilities before us. Whether influenced by the rabbit or not, the following day we put down a binder on the property and a few months later it was ours.

Since that first memorable trip from the Upper West Side of Manhattan to the wilds of western Orange County, we have driven countless times down the same tree-lined driveway; and, as often as not, especially in the dark of night, a rabbit jumps out to greet us.

During our first dozen or so years on the farm, the land on both sides of the driveway was hospitable to rabbits. In the beginning there was cow pasture along the north side. But under our tenure, absent the cows, this strip of land evolved into really fine rabbit country: lots of low bushes and scrappy young trees, thickets of wild rose and pleasant grassy clearings, ample food for a family of rabbits, and a wealth of good hiding places should danger threaten.

On the south side of the driveway, behind a rusted barbed wire fence, was our neighbors' property—several undulating acres of grassy open land peppered with shrubs and the occasional tree. The neighbors, an older couple, lived in a run-down house out front by the main road. To supplement their income, they bred Doberman pinschers. There were always a half-dozen distinctly unfriendly dogs tied up in their backyard. As soon as you got within fifty paces of these beasts they made it clear, by ferocious barking and neck-wrenching, that to advance further in their direction would be an act of folly. I'm sure the rabbits drew a similar conclusion.

About ten years ago, the neighbors' property underwent a transformation that did not favor the rabbits. It was sold to a developer. The neighbors moved out, taking their angry dogs with them. Soon bulldozers, backhoes, and construction teams moved in. The old house received a new roof, new siding, and new occupants, and, in a very short period of time, three additional homes appeared on the rolling land where the rabbits once had run.

Along with these homes came lawns, wells, septic systems, and numerous vehicles, including snowmobiles and ATVs. The change was dramatic. In the space of just a few months human society supplanted the society of rabbits. Now, when we encounter them on the driveway, the rabbits have less reason to hesitate. They know there's just one good direction to run in.

The driveway rabbits are a resourceful lot, and they soon adapted to the changes thrust upon them. While their numbers may be fewer now, due to the shrinkage of their territory, their population over the last few years has remained viable and remarkably stable. I would guess there are five or six of them living along the north side of the driveway at any one time.

Rabbits are not generally credited with high intelligence. I wonder, though, if they possess some knowledge or instinct that enables them to adjust their numbers as circumstances change—as space and resources become less available, for example. It is widely known how fast they can breed; a female eastern cottontail (which I believe these rabbits to be) can give birth to some twenty bunnies in a single year. If there were nothing to slow them down they might assume the title of "world's most populous mammal," which I would guess my own species can currently lay claim to.

Do they breed less when resources are scarce? Do they tolerate only a certain number of their kind on a given patch of land? (Owners of pet rabbits will tell you that two males in a cage together will fight until one or the other is dead.) Perhaps the cold winters curb their numbers, when the ground is covered with snow and food is in short supply. Or is predation the great balancing force? Are the young ones simply driven out of the nest and mostly gobbled up by foxes, coyotes, hawks, and other assorted flesh-eaters? Certainly our cats are happy to oblige in this regard—there's nothing they like better than to start the day with a breakfast of young rabbit.

Probably all of the above factors, and some others besides, make up the rabbit equation. Whatever the answer, whatever strategies are employed by or against them, there's little doubt in my mind that the rabbits who live along our driveway have arrived at a state of equilibrium with respect to the resources at their disposal. There's something deeply pleasing about this.

Unfortunately, the same cannot be said of rabbits in some other parts of the world. In New Zealand, where I grew up, for example, rabbits are something of a scourge upon the land. Introduced by English settlers in the nineteenth century, rabbits found the mild climate, absence of predators, and abundance of food greatly to their liking. Their habitat expanded as the settlers logged and burned the native forest to create grassy hills for grazing sheep. The rabbit population exploded into the many millions. Soon they were competing with the sheep farmers for grass and, along with the farmers, upsetting the native ecology. As a boy, I spent many a summer afternoon on my Uncle Rodger's farm hunting rabbits with an old single-shot .22 rifle.

Now, more than fifty years later, on my own farm in a different hemisphere, my feelings about rabbits are ambivalent. Because their numbers are modest and they legitimately belong to a larger community of plants and animals, I have come to regard the driveway rabbits as a group unto themselves, with certain basic and inalienable rights. They occupy a transitional or buffer zone on the farm and do not compete with me for resources. I don't go out of my way to promote their interests, but nor do I seek to cause them harm.

The same cannot be said of my relations with the rabbits that live in the hedgerows along the edges of our vegetable fields. While there is no shortage of grass and other foods palatable to them, these rabbits have acquired a taste for young seedlings of broccoli, collards, lettuce, and several other vegetables. When no one is watching, they nibble off the plants' first tender leaves, causing them to become stunted and unproductive, if not to expire outright.

Usually a few times each season my exasperation reaches such a level that I take down my shotgun and go out at dusk, bent on revenge. For a short while, the old hunting gene is reactivated. My senses sharpen; my pace becomes measured and stealthy. I am

When no one is watching, they nibble off the plants' first tender leaves.

Driveway Rabbit

possessed of a strange primal clarity. The first rabbit that comes within range doesn't have much of a chance.

It is with a mixture of righteous satisfaction and the beginnings of a vague guilt that I pick up the small limp body and take it back to the dogs or perhaps our own kitchen.

With my superior human intelligence, my predatory instinct, and my twelve-gauge shotgun, I have such surpassing power over these small creatures. The question then becomes, how many should I kill? Do I try to get rid of all of them, or just a few? And what about the driveway rabbits? When their numbers increase beyond the carrying capacity of their patch, surely some of them migrate down to the vegetable fields. Should I think about eliminating them, too? That's when I realize that I'm going too far.

Rabbits are not the only animals that threaten the vegetables we grow; nor are they the most troublesome. Woodchucks and deer are far worse. Chipmunks, field mice, and meadow voles can also do significant damage. For the past couple of years we've had so many carrots nibbled by voles that we've started separating them and selling them at a discount.

The voles leave a little ring of tooth marks around the tops of the carrots where they protrude out of the ground. Sometimes they even dig the soil away and dine at a deeper level. We promote these carrots as "pre-eaten" or "taste-tested by voles" and may at times have made the somewhat dubious claim that voles select only the sweeter carrots to nibble on. Last fall, one of my longtime customers coined the term "vole-uptuous" to describe this new line of produce.

As my own species continues to expand its population and tighten its grip over the earth, often with scant regard for other forms of life, I wonder about the role I play as a small farmer, always turning the land to human needs and my own profit. How far should I go? And to what extent is it really "my" land? When is the damage to the vegetables too great? At what point do I exert lethal force against competitors—and will it make much difference anyway? Or should I be content to share this rough old farm with its many nonhuman inhabitants, regardless of the cost? How little my customers know of the sometimes troubling issues that lie behind each bite of carrot or broccoli.

The answers to these questions are never easy or simple. For me, they change with mood and circumstance. Some things, though, are nonnegotiable: Among them are the driveway rabbits. I hope never to see the day when they are gone.

SUSTAINABLE vs. ORGANIC— WHO LOSES?

Now that the word *organic* is somewhat tarnished by its association with the federal government, I find myself using the terms *sustainable* and *local agriculture* more than I used to. It's not that *organic* has become a dirty word—it still means farming without synthetic chemicals and that's still a good thing. But for many small farmers and an increasing number of consumers, the word lost some of its luster soon after it received the federal and corporate embrace.

Ever since the USDA's definition of organic became the law of the land, farmers wanting to use the word to describe their products and processes have had to conform to federal standards and submit to a complicated and costly certification process. To do otherwise may invite hefty fines.

Until recently, most organic farming was more or less synonymous with small-scale, local agriculture with a distinct ecological and philosophical bent. Today the stakes are higher, and many of the players are larger: Agribusiness interests have established operations in the United States and are tapping into cheap labor in other countries to produce crops that conform to the USDA's organic regulations. To carry the green-and-white organic label, it is not necessary that crops be grown or produced in this country. A recent USDA study estimated that, in 2002, organic food imports into the United States from South America, China, and elsewhere reached a value of $1.5 billion. In the same year, U.S. organic exports were in the neighborhood of $125 million. Thus we already have a large organic trade imbalance, which does not bode well for American farmers. And consumers might note that in a world of international contract growing, mass marketing, and thousands of miles between a food's place of origin and its retail destination, ecological, philosophical, and even nutritional considerations can be easily swept aside.

Industrial agriculture appears to be happy to stay with its chemical habit on the one hand and, on the other, invest in the growing organic market. It sees no contradiction—after all, both offer economic opportunities. In particular, the processed and packaged organic food industry has become irresistibly profitable to the corporate food giants. The shelves of health food stores are lined with packaged products carrying the organic label along with fairly high price tags. Supermarkets, too, are expanding their "natural foods" aisles and stocking them heavily with processed and packaged organic items. Just how much of an improvement these products are over their conventional cousins is not clear. What does seem clear, though, is that the food industry would like to eliminate as many of the differences between the two as possible.

The Organic Trade Association (OTA) and corporate lobbyists have been working hard to persuade Congress to relax and reshape the national organic standards to better meet their clients' needs. The big food players want to cash in on the lucrative organic label with as little modification to their customary practices as possible. The political leadership of the USDA seems not to have a problem with this. It is generally their habit to accommodate large agribusiness interests. But, on the other side, there is impassioned public opposition to any weakening of the December 2000 national organic standards. How long individuals, public interest groups, and small farmers can fight off these forces of dilution is anybody's guess.

Cognizant of these undermining corporate influences, some small farmers, formerly certified organic by nonprofit, farmer-run organizations, have chosen not to participate in the National Organic Program. They are willing to forgo the use of the word organic altogether, even though they may have worked for years to build the credibility and market share that organic agriculture now enjoys. Instead they are turning to words like *natural, biological, regenerative, eco-local,* and *sustainable* to describe what they do.

Their reasons are not difficult to understand. Some are unwilling to submit to another federal bureaucracy; some cannot afford, or do not wish, to pay the fees; some are opposed to a few of the arbitrary federal requirements they believe do not make sense; and almost all do not wish to lie in the same bed with agribusiness.

I'm in my eighth year of certification under the federal program. Prior to that, I was certified by the Northeast Organic Farming Association of New York (NOFA-NY) for fifteen years. NOFA is still my certifying agent, but they have undergone a name change to NOFA-NY Certified Organic, LLC, and they are no longer able to write their own rules or standards. Today, they follow the national standards established by the USDA, though they do have some leeway in interpreting them. I've made the decision to stay with NOFA and the federal program, but I am uneasy with it and am making no long-term commitment. That's why words like *sustainable* have increasing interest for me.

But what exactly does *sustainable agriculture* mean? It is a broad concept, to be sure, and probably no two people in a room would come up with precisely the same definition, though perhaps most would agree that for agriculture to be sustainable it has to remain viable for future generations, not just for ourselves. That's not a bad place to start.

To me, sustainable agriculture has a lot to do with a sustainable planet. It is a way of farming and living that assumes we are going to farm and live on this planet forever, and be happy with that. It does not assume that technology will in time solve the problems that we create today (though in some cases it might). It views the farm as a complex and intricate ecosystem that supports and relies on the involvement of many organisms—from the micro to the macro, from the fungi and bacteria that inhabit the soil to the plants and animals that live on the land to the birds and insects that fly above it.

Sustainable agriculture is about more than just the quality of food and fiber that are raised, harvested, and sold—it's about the health of the land being farmed and ultimately the health of all other land. Presumptuous though it may be, when things are going well, I look at our farm as one small attempt to create a model for a sustainable planet.

Sustainability is about having more small farms and fewer big ones, more farms run by individuals, families, and partnerships and fewer farms run by publicly held corporations. Every farm, like every family, is unique. The small farmer who lives on his or her land will learn over time what is best for it. He or she will likely develop a fondness for the land and a sense of belonging, and through these will come a sense of stewardship. Bedrock issues—such as fostering diversity, controlling erosion, and maintaining a dynamic living soil—will be natural priorities.

Sustainability is about having more food produced and consumed locally, or at least regionally. When a farm is part of a community, be it an organic farm or otherwise, it belongs, in a sense, not just to the farmer, but to the community as a whole. It enriches the community by providing open space, wildlife habitat, and scenic views. And it keeps alive a connection to one of the most elemental of

Sustainable agriculture has a lot to do with a sustainable planet.

human activities—the cultivation of food. The small, well-managed farm, functioning within a community, is both an antidote to and the antithesis of global corporate power.

Mere adherence to a set of federal rules does not ensure sustainability. The management and stockholders of an agribusiness corporation (even one producing organic food) are unlikely to have a personal connection to the farmland in which they are invested—their primary and often only concern is for the value of their shares and the dividends they might receive. For them, the long-term health of the land is likely to be of no consequence. The land may be viewed as little more than an extractive resource or a commodity to be sold at any time, for any purpose, if the price is right. In this context, organic becomes just another marketing opportunity, with the heart and original intent gone from it.

Good farming should never be exclusively focused on short-term profit, especially if it is at the expense of long-term health. Unfortunately, our society places great importance on the former and has little time or patience for the latter. Farmers, like everyone else, are caught in the profit imperative.

And yet, all agriculture, sustainable or otherwise, is and must be concerned with profitability. If a farmer cannot make a living wage, then the business is not sustainable and all the good intentions and high-minded ideals will be of no use. This is an enormous problem. Many small farms fail each year because their owners cannot eke out a living from their land, no matter how hard they try. This remains painfully true for dairy farmers in the Northeast, who often receive less for their milk than they did twenty-five years ago. Along the food chain—from the commodities trader to the wholesaler to the trucker to the retailer—everyone makes a living. But the farmer—the primary producer, the one entrusted with the health and productivity of the land and the health of the food we eat—often is left with a pittance, or nothing at all.

If true sustainability is our goal, an almost revolutionary change in policy and attitude is needed. Farmers alone cannot build and maintain a sustainable agriculture. Supportive governmental

Sustainability is about having more food produced and consumed locally, or at least regionally.

policies would be of great assistance, but at this point, there are few among us who expect our government and its corporate paymasters to lend a helping hand.

If there is to be a real shift in thinking, a real movement toward a vital and durable food system for all Americans, it will be driven by the consuming public. To this end, the small farmer must become an educator, even an apostle. We must give the public the healthiest, freshest, and best-tasting food we can. We must also be willing, if we are suited to the task, to tell consumers how their food is produced and to talk about the importance of a sound, accountable, local food system. The public should understand why it is in their interest that small farms survive and prosper.

I would ask that consumers learn to look beyond the food they are eating and its effect on their bodies and consider also the well-being of the land that provides the food. I would ask that they, too, develop a sense of stewardship toward the farms and fields that nourish them, and the wide earth beyond. It is vital that consumers and farmers come together to forge a partnership in sustainability.

More environmental organizations might also take up the banner of sustainable agriculture. Some, such as the American Farmland Trust, the National Campaign for Sustainable Agriculture, and the Regional Farm and Food Project already have. But many major environmental advocacy groups seem reluctant to look too closely at our food system. Preserving rain forest, saving endangered species, and keeping our oceans and waterways clean are worthy goals that I wholeheartedly support. But if we are willing to disregard and despoil the land that produces the food we put into our bodies, how grounded is our commitment to a livable planet? When there is a call to war, can the rain forest still hold our interest? How long will the Arctic National Wildlife Refuge remain closed to Texas oilmen? Creating an earth-friendly and truly sustainable agriculture could become the most basic of environmental goals—one from which all other efforts to preserve our planet gain resonance.

The small, well-managed farm, functioning within a community, is both an antidote to and the antithesis of global corporate power.

INNER SANCTUM— an office with a view

have just moved into my new office, an ample and sparkling 275 square feet of finely crafted space. With windows on three sides, a handsome oak floor, a high cedar-wood ceiling, two large closets, a built-in work station, and a window seat complete with bookshelves, here is an office fit for a prince, or at least a gentleman farmer, though I have yet to attain either status.

"Why does a farmer need such an office?" some might ask. Are not the open fields his place of work? Is not the sky overhead his rightful canopy? Surely a corner of the tractor shed or barn would suffice for the few seed catalogs and papers needed?

To such naïveté, I would answer that a small farmer who wishes to survive as such must wear many hats—and some of these hats can only be worn indoors. A farmer cannot afford to hire a cast of professional bookkeepers, accountants, secretaries, retailers. He or she must perform these functions him- or herself. There are myriad records to be kept, forms to be filed, applications to be filled out, and bills to be paid.

Federal and state governments do not discriminate on the basis of size. Most of the rules and requirements that pertain to large corporations are equally applicable to the small entrepreneur who employs workers beyond his immediate family. For these reasons, I feel entirely justified in having a substantial and well-endowed office. The eight-by-nine-foot room in which I have struggled to maintain order for these past eighteen years is no longer adequate.

The new office took four months to build. This period was preceded by a few months of planning, in which several options were considered, sketches made, friends consulted, architect and contractor engaged, blueprints drawn, and, finally, a building permit obtained. In many ways the planning process was more taxing than the construction itself.

Our house, which originally served as tenant quarters on a larger farm, was built at least 125 years ago. Back then it had just three rooms on two floors, providing a total living space of about 800 square feet. Over time, subsequent occupants added a separate kitchen, a small bedroom and bathroom, and, most recently, a sunny front room built on a concrete pad, all of which brought the total living area up to around 1,400 square feet. The new office sits atop the front room on the concrete pad but is cantilevered out two feet beyond its base.

One Sunday afternoon ten years ago, an older couple drove down our driveway, introduced themselves as Al and Myrtle, and said they had lived on the farm from the late 1940s through to the mid-1960s. They spoke fondly of their time here. It was their first house and their first farm. They had just got married and were in the spring of their youth. Al milked cows and Myrtle raised their children. It was a time of relative prosperity for farmers. And it was a time of transition, as horse traction gave way to tractors and new chemicals seemed to increase productivity and make farming easier.

When they came to the farm, Al and Myrtle used a shallow well and a two-seater outhouse. They lived without hot running water for the first three years and it was eleven years before they enjoyed a flush toilet. Myrtle recalled, with a smile, that for Christmas one year, Al bought her a portable bathtub from Sears, which they set up in the kitchen. The current bathroom was built around 1960 and is part of a simple lean-to addition to the house.

Since Al and Myrtle's time, other families have lived and worked here, each leaving behind memories and a part of its own history. When the mood is right, I like to think, one can sense these past lives in the fields and groves of the land, along the old stone walls, and in the aging rooms of the house. Sometimes, clues to the past are quite overt. On the vertical molding inside the pantry door, we found a progression of growth lines of one family's children—neat marks on the molding with notations indicating date, height in inches, and the name of each child.

During our years on the farm, my wife, Flavia, and I have embarked upon several home-improvement projects, some in the category of much-needed maintenance, others with more aesthetic goals in mind. We've insulated where there was no insulation, replaced crumbling asphalt siding, built a sturdier roof, installed new windows, raised ceilings, and performed renovations to kitchen, bedroom, and bathroom.

Each of these improvements was preceded by some fretting and indecision but, upon completion, was deemed well worth the effort and expense. None, however, increased our living area.

The new office, I am pleased to say, represents the first addition of space to occur during our tenure. It is our contribution to the evolving mosaic of the house. For this reason (and because it is such a well-constructed room), it brings pleasure and delight of a whole new order. While it has certainly caused some disruption (for example, the roof over the front room had to be removed entirely and the room wrapped in plastic, like a Christo sculpture, to keep the rain out), it has been an exciting process to witness and participate in. Along the way, many choices and sometimes difficult decisions had to be made, but at the end of each day there was always visible progress to enjoy.

It would not be untrue, nor would it reflect unfairly on the remainder of the house, to say that the new office is, without doubt, the finest room under this one roof. It receives natural but mostly filtered light from three sides. It has pleasant views. It is a well-proportioned rectangle with an alcove for a window seat and a sloping ceiling in the front that reflects the pitch of the new roof. Because it is upstairs and in one corner of the house, it has the quality of a retreat, a place of quiet and contemplation.

Of overriding significance is the fact that this new room was designed and built by a true artisan, a man by the name of Tom Berg who enjoys his work and takes pride in it. His creative stamp rests lightly on every feature of the room. All who enjoy it over time will remain, in no small way, indebted to him for his care, his attention to detail, and his love of form.

A good part of the room's special character arises from the selection of materials from which it is fashioned, especially the several different kinds of wood.

In one corner is a broad, hand-hewn post of the original house; rough adze marks are clearly visible, as are the dowels or pins that tie a sturdy horizontal beam and wind brace to the corner post. Nails that were used in this corner are square-headed and appear to be hand-forged. Growth rings on one of the boards of original lumber—probably hemlock—suggest a tree that may have begun its life before European settlers arrived in North America. You can tell from the spaces between the rings when the tree had a good year of growth and when there wasn't enough rain.

Under the window seat there are three large drawers, partly fashioned from old-growth oak salvaged from stanchions once used to confine cows in the barn. We dismantled the stanchions fifteen years ago to open up more space; I put aside some of the wood,

There are myriad records to be kept, forms to be filed, applications to be filled out, and bills to be paid.

steeped in the history of cows, hoping that one day I would find a good use for it. This beautiful wood is dense, heavy, and finely grained. It is so hard that nails, unless given predrilled holes, will not penetrate it.

The ceiling of the new room was the subject of extended debate. Should we simply paint the Sheetrock that Tom had already put up, or cover it with some kind of wood? For a while, shiplap pine was at the top of the list. Then, in the upper barn, I uncovered some tongue-and-groove cedar, left over from an earlier project. It was exterior-grade lumber, full of knots, but attractive nonetheless. This well-cured wood has found a new home on the office ceiling. Stained a pickled white, it reflects a soft light.

The office floor is made of two-and-a-quarter-inch-wide strips of red and white oak of varying lengths and hues. The wood was salvaged from Tom Berg's scrap pile and passed on to us in exchange for a month's supply of fresh vegetables. Around the perimeter of the floor, about a foot from the edge, is a narrow band of black walnut sandwiched between two strips of cedar. This border functions like a frame, setting off the variegated pattern of oak within. Both walnut and cedar were taken from old boards found in the barn.

Near the center of the room is an ancient walnut desk with very deep drawers and a surface that appears to be one large slab of wood. It is a desk that has seen much use but has maintained itself well. It is big and weighs 225 pounds when empty, which is perhaps why others passed it by. My wife bought it at a neighbor's yard sale for $10, then sent me over to pick it up.

As winter approaches, I look forward to spending more time in my new office. I look forward to enjoying this fine room and becoming more intimately acquainted with its many aspects. The various parts, so well assembled into a whole, seem to hold memories of other lives and other times. This quality, this ethereal door into the past, will, I believe, be a source of sustenance as I grow older, and perhaps will help me to knit my own small life into a larger fabric.

A REVERSAL of FORTUNE

Lazarus, the rooster, has suffered a reversal of fortune and I must take some of the blame. While still an active bird, Lazarus is undoubtedly entering his mature years. With fifteen hens to look after, he has, of late, not always been able to fully meet his conjugal obligations. I felt the time had come to introduce into the flock a younger male who might gradually take over as Lazarus slipped into his dotage.

I went to my Greenmarket friend Don Lewis, who has generously provided me with birds in the past. Might he have a young rooster to spare, I asked. "Yes," said Don with a grin, "I think I do."

*Jose—the
Ruler of the Roost*

We set up a transfer date, and a week later a new rooster rode back with my crew from the Wednesday market. In the dark of night I removed him from the cabbage box in which he had been confined for some thirty hours, noticing, as I did, the name "Jose" scrawled across the side of the box. I was a little surprised by his weight and girth. For a youngster, he was a big bird, with sturdy feet and well-developed spurs.

Taking a firm grip around his legs, I transported him into the coop and set him on the roosting bar between two sleeping hens. For a moment or two there was some nervous clucking and shuffling, then all was quiet again.

The next morning I went out to check on the newcomer, expecting that he would be keeping a discreet distance between himself and Lazarus. I immediately spotted him, surrounded by a group of hens. In the daylight

While still an active bird, Lazarus is undoubtedly entering his mature years. With fifteen hens to look after, he has, of late, not always been able to fully meet his conjugal obligations.

he was quite spectacular, a dark and golden bird with long iridescent-blue neck feathers, almost like a mane. But where was Lazarus? I looked all over, wondering if he might have escaped from the coop. When I found him in a back corner, his head was buried in a wild rosebush and some of his tail feathers were a little askew.

I nudged him gently with my foot. When he turned to look at me, I was shocked by what I saw and felt a sudden surge of guilt. His formerly handsome face was bleeding and torn, his once proud comb flattened against his head. One of his eyes seemed displaced from its usual location, and the fighting spur on his right leg was broken and bloody. Wretched and bedraggled, he limped away, looking for another place to hide. It crossed my mind right then that I might be doing him a favor were I to take an ax and remove his battered head completely. But I did not have the heart to do that.

I am no stranger to rivalry between roosters, but have never before witnessed such a rapid and resounding reversal of fortune. Lazarus, who should have had the advantage, was literally trounced before breakfast. How humiliating for him.

Now, a few weeks later, there is relative calm in the coop. Lazarus's wounds have mostly healed, but he has lost his air of confidence and seems to have shrunk in size. Jose now rules the roost but appears willing to tolerate Lazarus so long as he stays in the shadows and does not cast an amorous eye on any of the hens. To do so could be fatal.

THE UNWEEDED GARDEN

"'T is an *unweeded* garden, That grows to seed," noted Hamlet, and I couldn't agree with him more, though I doubt his experience in weed control is equal to mine. For this reason I must take slight offense at any implied criticism in his remark.

Each year we spend countless hours combating weeds so that our crops may prosper. Usually, they do, more or less. But always some weeds, and often many weeds, get the better of us and persist, despite our efforts to eradicate them. Sure enough, they go to seed (or "grow to seed," if you prefer), and in so doing create generations of future weeds that will require further countless hours of work on our part in seasons to come. It is an ongoing struggle, rather like life itself. I do not expect ever to prevail over the weeds. And perhaps it is just as well.

Weeds have successfully colonized the earth. They surely predate us and our modern vegetables, most of which, if truth be told, have very weedy ancestors. What is a weed, after all, but a plant that we have no immediate use for and that gets in our way?

If there is land available and suitable to their needs, especially tilled or disturbed land, weeds will occupy it far more aggressively and effectively than any domesticated vegetable that I might insert in the ground. And this is why weeds are such a problem. Simply put, they are much better at what they do than the plants our customers expect us to grow. If left alone, they would outcompete our vegetables every day of the week.

Individual species of weeds have specific preferences with regard to soil acidity, fertility, texture, drainage, and so on. The knowledgeable farmer can deduce much about the condition of his soil, including its excesses and deficiencies, from the types of weeds that inhabit it. Some weeds flourish in acidic soils, some in alkaline soils, others in heavy, compacted, or poorly drained soils, and still others in fertile, well-drained land.

The weeds that frequent my farm in large numbers—chickweed, quickweed, pigweed, lamb's-quarters, burdock—are species that favor fertile ground. While pleasing on one level, this hardly endears them to me, since they grow with such promiscuous abandon and inevitably deprive my vegetables of needed nutrients, moisture, and space.

In terms of life span and reproductive strategy, weeds can be divided into three groups: annuals, biennials, and perennials. Most of our problem weeds are annuals that live for just one season. Usually they do not form deep roots and therefore may be dispatched with a hoe if one goes to them in a timely fashion. But annuals grow rapidly, are quick to set seed, and are prodigious multipliers.

The gentle and harmless-sounding lamb's-quarters, for example, can grow six feet tall and produce fifty thousand seeds or more before succumbing to a fall frost. In their youthful phase, a scattering of little lamb's-quarters can appear quite attractive and benign; if left alone for a month or two, however, the resulting forest of plants is a sobering sight for a farmer to behold. A single red-root pigweed plant, also an annual, if left to its own devices, can produce several hundred thousand seeds in one short season of growth.

As is the case with lamb's-quarters and most other weeds, the seeds may be dispersed by wind, birds, animals, and even humans. Or they simply drop into the soil at the base of the mother plant, where they can remain viable for many years, waiting only for the right combination of temperature, moisture, and light to trigger their germination.

Biennial weeds, such as burdock and Queen Anne's lace, live for two seasons and set seed in their second year. They grow more slowly than annuals but, once established, are harder to eliminate. With some effort, one might dig up most of a burdock plant; any pieces of root that remain in the ground, however, will very likely regenerate and come back to defy you.

Perennial weeds can live for three years or more. They tend to have deep, tenacious roots that are almost impossible to pull out of the ground in their entirety. Like biennials, they often reproduce both by seed and vegetatively. Getting rid of them is a serious challenge. Dandelion and Canada thistle are common perennials. We have plenty of dandelion in our lawn, where I find it quite attractive when it first blooms in May. I am less enamored of dandelion in our fields.

Stirrup Wheel Hoe

Carrot Weeder

While it would be less than chivalrous not to have a certain grudging respect for most weeds—on account of their toughness, their tenacity, their lust for life, and sometimes even their untamed beauty—I, like other farmers, must take action against them if I wish to have any product to sell.

Conventional farmers may use synthetic chemical herbicides for weed control. Organic growers do not have this option. On our farm the ongoing campaign against weeds is conducted on several fronts and with an assortment of weapons. One of my favorites, especially for dealing with young annuals, is the wheel hoe. This low-tech device has two long handles with a wheel in between. A few inches behind the wheel is a blade, often shaped like a stirrup. The user simply walks between parallel rows of vegetables, pushing the wheel hoe and allowing the blade to slice just below the soil surface, effectively decapitating or dislodging any weeds in its path.

An experienced wheel hoe operator might make his or her way through a half-acre field of vegetables in two or three hours, assuming the weeds are of manageable size. Weeds growing between plants within the rows must then be dealt with using a long-handled hoe, a more time-consuming process.

Tractor cultivation involves the use of a tractor to pull a cultivator, rotary hoe, or some other weed-destroying implement through the field. Weeds are uprooted, buried, chopped up, or otherwise undone. Tractor cultivation makes the most sense when one is dealing with large, uniformly planted areas with enough space for the tractor to get in and work efficiently, without doing collateral damage to the planted rows. On our farm, winter squash and pumpkins (before their vines start sprawling) are good candidates for this kind of mechanical weed control.

At the other end of the spectrum of weeding methods is the age-old technique of hand-pulling. We do plenty of this, as well as working on our hands and knees with small hand hoes to control weeds around tiny and vulnerable plants such as young onions or carrots.

Wheel hoeing, hand hoeing, tractor cultivation, and the time-tested hand-pulling technique are all best performed during the earlier part of a dry, sunny day, preferably when there is a little breeze in the air. Many uprooted weeds have a remarkable ability to reroot themselves and start over again, which is far more likely to happen in damp, overcast conditions than on a clear, windy day. The objective in weeding is to have your victim's roots dry out as quickly as possible.

Weeds also may be controlled through the use of mulches, which act as a physical barrier preventing germination and/or plant growth. Organic mulches include straw, wood shavings, chopped leaves, and even newsprint. In time, these mulches break down and blend into the soil, enhancing organic matter content. If this happens during the growing season, more layers of mulch may need to be added.

Black plastic is an inorganic mulch. It is good at blocking weed growth but must be removed and disposed of at the end of each growing season. More long-lasting synthetic woven mulches or weed barriers are also available and can be useful in a perennial garden. There are some biodegradable plastics on the market, but they have a tendency to break down in the field before you want them to, especially in a wet season.

The farmer (or gardener) who relies on plastic mulch for weed control must poke holes in the plastic and then plant through them into the soil. Planting through mulch, whether plastic or organic, is more time consuming than planting into bare soil, but saves weeding time later on.

Some crops, notably garlic, will grow up through wood shavings or straw if it is not too thick. This means the mulch can be spread immediately after planting and one can avoid the step of making an individual hole for each plant.

We use a fair amount of wood shavings and straw and some black plastic. The organic mulches work best with crops that prefer cool, moist conditions. Mints thrive when mulched with straw and kept moist, as do most greens. Black plastic works well with heat-loving plants such as tomatoes, peppers, and basil.

The most martial weapon in my weed-control arsenal is undoubtedly the flame weeder. This diabolical device shoots out a searing flame that either burns weeds to a crisp or desiccates them on the spot. A tank of propane is strapped to a pack frame and carried on one's back; a three-foot wand with a trigger is carried in one hand and directed at the target weeds. When working with a flame weeder, it is advisable to wear long pants and sturdy footwear—bare feet would be a big mistake.

Red-Cored Chantenay Carrot

The flame weeder works well with direct-seeded crops that may be slow to germinate and put on growth—carrots and parsley are two examples. (These and a few other crops we grow, such as beans and turnips, will not consent to transplanting. They are always directly seeded in the field with a small walk-behind hand seeder.) When given optimum soil temperature and moisture, carrots can germinate in a week. Lamb's-quarters, pigweed, and several other interlopers might germinate in four or five days. The farmer who understands this will go to work with his flame weeder on day five or six. The tiny, just germinated weeds, almost invisible, quickly succumb to the withering heat. A couple of days later (if all goes well), the carrots will emerge and can then enjoy several competition-free days. With the flame weeder, timing is everything.

One would be hard-pressed to name any native plant or animal that does not have some value in the larger scheme of things. Weeds are no exception to this fundamental rule of nature. While they surely increase my workload and are the source of much exasperation, I am well aware that without them the world would be a different and deficient place. Like the hunter who quietly loves and respects his quarry, I must now look beyond my adversarial relationship with weeds and offer a brief defense of them. In truth, they perform many valuable functions.

- Weeds provide important habitat for insects, birds, and animals. They offer them food, shelter, pollen, and places to hide.

- Like cover crops, weeds prevent erosion by holding the soil together, especially during periods of heavy rain or in spring thaw and snowmelt. Also, like cover crops, they contribute organic matter to the soil when they die.

- Weeds scavenge and mine for minerals and nutrients deep in the subsoil where the roots of vegetables are unlikely to go. In so doing, they open up and aerate the lower reaches of the soil. The deep-captured nutrients enter the weeds' vascular systems and bodies; when the weeds die and decompose, these nutrients are released into the upper soil layers, where they become available to crop plants. Some nutrients and minerals may leach out of the soil altogether if they are not incorporated into the body tissue of weeds.

Praying Mantis

- Certain weeds have medicinal value and have been used therapeutically through the ages; others are nutritious foods. Many of my problem weeds are highly edible; some we even harvest and bring to market. Lamb's-quarters has a wild and nutty flavor and is packed with food value. When small it can be steamed, stir-fried, or added to a salad or soup. Purslane, another weed found on our farm, has succulent stems and leaves that contain plenty of iron, as well as calcium, phosphorus, and vitamins A and C. Even red-root pigweed, a member of the amaranth family, is edible when small. Young chickweed makes a good addition to a salad and is coveted by our flock of hens. Burdock root, after considerable preparation, is excellent in soups, stews, and sautés.

- I have already mentioned the value of weeds as indicator species. More surely than any vegetable, they can tell us about the condition of our soil and what measures might be taken to improve it.

- Above all, weeds, like other wild things, are part of the intricate web of life that often transcends our limited human interest. They represent the triumph of diversity over monoculture, difference over sameness. They are a vast genetic storehouse of plant knowledge and resilience that reaches back through earthly time.

The Jesuit English poet Gerard Manley Hopkins understood the vital nature of weeds and perhaps even saw the hand of God in them. (Though I suspect that, like Hamlet, he wasn't much of a gardener.) In his poem "Inversnaid" he wrote:

> *What would the world be, once bereft*
> *Of wet and of wildness? Let them be left,*
> *O let them be left, wildness and wet;*
> *Long live the weeds and the wilderness yet.*

Walk-Behind Seeder

FARMS
on the BLOCK

Every year in the United States, a million or more acres of productive agricultural land are converted into housing developments, office buildings, roads, malls, parking lots—the trappings of modern society on the move. It's a one-way street. Once the orchard is cut down, the last cow is sold at auction, and the bulldozers move in, there's no going back. The farm is gone forever, like an extinct bird. Only a wistful memory remains for those who knew it. The newcomers, with their landscaped gardens and overfertilized lawns, are oblivious to the robust life that once flourished beneath their driveways.

Many think this loss of farmland, this human encroachment, this suburban sprawl is inevitable, something like manifest destiny. Within the real estate industry, and perhaps in society at large, it is axiomatic that the "highest and best use" of a piece of land is the one that brings the greatest economic return. These days, small farms are seldom on the right side of that equation unless they are in remote areas with no development pressure.

As we proceed into this new millennium, a nation and people at such a pinnacle of affluence and power, I would like to make a case for considering land use and progress a little differently. We are no longer a frontier nation. A hundred and fifty years ago there was a surplus of fertile land and not enough people to cover it; today there are many of us and good farmland is a diminishing resource. Perhaps the time has come to reckon its value in other than purely monetary terms.

Admittedly, I have a personal stake in this matter. I have always liked farmland and open space, and it pains me to see them disappear. And they are disappearing—all around me: A quarter mile to our North, a fourteen-home subdivision was recently completed. On our eastern border, three new homes have been constructed. Development pressure, of course, is not confined to my immediate neighborhood. In "Farming on the Edge," the American Farmland Trust recently identified the Hudson Valley as one of the nation's top ten most-threatened farming regions.

On the eighty-eight-acre farm on which I live with my wife and seven or eight seasonal workers, there are only about thirty acres of land that are actually tillable (i.e., can be plowed and planted to crops like vegetables, corn, or hay). Another twenty-five acres or so is pasture—not well drained enough for field crops, but suitable for growing grass. Each summer this area is home to eight or ten cows waiting to give birth. The balance of the land is mixed woods with ridges and rocky outcrops and a few wet spots. To me this is the wild heart of the farm.

Like many small farms, ours has a rich diversity of life. Its patchwork of fields, woods, and wet areas (typical of much of the

Perhaps the time has come to reckon [the value of land] in other than purely monetary terms.

*Baltimore Oriole
Fledgling*

Northeast), makes ideal wildlife habitat—you're likely to find more species of plants, animals, and birds around small farms than in large areas of mature forest. And, organic management of the land is a big plus. A recent British study found that organic farms in England host a higher diversity and larger numbers of plants and animals than their chemically sprayed counterparts. (The government-funded study looked at 180 farms over five years and was conducted by scientists from Oxford University, the British Trust for Ornithology, and the Centre for Ecology and Hydrology.) Several years back, a farmworker at our place—who also happened to be an expert bird-watcher—identified seventy-three different species of birds on the farm during the course of one growing season. He gave me a copy of his "birding list" at the end of the year as a parting gift (it is reprinted in the appendix).

Squirrels, chipmunks, groundhogs, rabbits, opossums, raccoons, skunks, foxes, bats, coyote, and deer are just some of the more noticeable animals that regard the place as home. For the most part, they coexist with us in relative harmony—over the years I've come to think of them as having at least as much ownership of the land as we do. If it were just us and the vegetables it would be a poorer place. Today, at dawn, when I went out to relieve the dogs from their nightly deer-deterrent duties, a great blue heron glided down through the mist that lies over the pond on autumn mornings. It alighted on the shore near a cluster of cattails and stood there in total stillness and silence, waiting for a frog for breakfast.

When a school bus drives by a farm in an otherwise residential setting, the children witness something different: fields of corn,

grazing cows, rough hedgerows. Each of these contain aspects of nature and knowledge that are not apparent in the classroom or even the large backyard. The fertile land is more expansive and alive. It gives off a feeling of freedom and connects us to the vital underpinnings of our lives. We may take the open fields for granted or not register their significance, but they leave an impression, nonetheless, to counter the unreality of the TV screen and so much else.

Small, diversified farms can function with houses and schools around them and may even benefit by having a ready market for their products. And for local residents, knowing where the food they eat comes from and who grew or raised it adds a whole dimension of confidence and connection. This is what healthy communities are made of.

Opossum

How, then, do we prevent shortsighted market forces from obliterating all the farms in their path? What steps can we take to moderate development and keep productive farms in our midst and within reach of our great cities? There are no easy answers, but a few points can be made.

Most farmers will keep farming if they can make a living at it. But for many this is extremely difficult. Dairy farmers, for example, are trapped in an arcane national pricing system that fluctuates wildly and has something to do with the price of cheese in Eau Claire, Wisconsin. The large cheese processors in the land of lakes like to keep milk prices artificially low and generally manage to do so. Small dairy farmers invariably get the short end of the stick. At the supermarket, we'll often spend more for soda or bottled water than we do for milk. This is lopsided and unfair—milk is real food; soda is carbonated water and sugar with a whole lot of advertising behind it and not much else. Our policy makers and politicians and all of us should find a way to pay dairy farmers what they are worth.

The farm is gone forever, like an extinct bird. Only a wistful memory remains for those who knew it.

Agricultural and conservation easements and other protective strategies can be used to preserve farmland. New York State has a small farmland-protection program that pays farmers to give up the right to develop their land. If the farm is sold, all future owners must maintain it in agriculture or at least open space. Benefits accrue all around. The current owner is partially compensated for not selling his property to the highest bidder (usually a developer); a future farmer is able to purchase the land at a more affordable price that reflects the land's value as a farm, rather than for commercial development or subdivision; and the community wins by maintaining diversity and a rural atmosphere and by keeping a lid on property taxes. It's a fact that new development always calls for more infrastructure, bigger schools, and more services, which inevitably leads to higher taxes. On the property taxes issue alone, preserving farms is a good deal for local residents.

Unfortunately, New York State's farmland-protection program is small and woefully underfunded. Very few of those who apply are accepted. Farms with the best soils (namely, Class I) have a better chance. We grow most of our vegetables on Class II soils—the second best out of eight agricultural capability classes. These soils work fine for us but have plenty of rocks and occasionally suffer from less than perfect drainage. For these reasons, we are not likely to be accepted into the state's farmland-protection program. Twice we have applied to sell our development rights to the state, but in neither case did we get so much as a postcard acknowledging receipt of our lengthy application.

We need much more state funding. New Jersey, a fraction the size of New York, has committed $35 million per year for ten years to preserving farmland and another several hundred million dollars has been set aside to preserve nonagricultural open space. A more generous contribution from New York State could stimulate greater interest and participation at the local level and among private conservation groups.

Where development is inevitable, town and county planning and zoning boards need to work with developers to find creative solutions that preserve as much farmland as possible. Smaller lot sizes for new homes and clustered development are among the possible solutions. Lot sizes of two, three, and five acres—common where I live in Orange County—use up land fast. They separate neighbors from each other and leave them with large lawns to mow on the weekend.

We need diversity. We need small farms. And we need them among us.

Instead of breaking up a hundred-acre farm into twenty five-acre lots, why not cluster the twenty lots on ten or twenty acres and leave the remaining eighty or ninety acres protected and available for use by a farmer? Such alternative development plans have been used by conservation advocates for years to protect environmentally

Raccoon

sensitive land while allowing for growth—why not apply the same principles to protect farms, too? With good planning and a flexible, solution-oriented approach, there's no reason why we can't have new homes and farms side by side. Balance is the key.

In the last few decades, ecologists have helped us to understand that diversity in living systems is synonymous with health and resilience. At the same time, we are beginning to learn that monoculture—whether in the form of a vast corporate hog farm, ten thousand acres of subsidized soybeans, or hundreds of square miles of suburban sprawl interrupted only by strip malls—is not in our best interests. Such growth may satisfy a corporate agenda, but at the end of the day we know that this kind of growth is "weary, stale, flat, and unprofitable," to once again quote Hamlet, and it gives progress a bad name.

We need diversity. We need small farms. And we need them among us. Let's not keep living out the lines in the Joni Mitchell song that tell of paradise being paved with a parking lot.

It's a Long Road
to a Tomato
194

THE HEART of WINTER

(January 2003)

nother cold night, close to zero degrees, with a fierce wind blowing. How do the sparrows and wrens survive in this, clutching the limbs of trees? And now that the sun is up, it's not much better—maybe ten degrees, and the wind still raging across the frozen land. This is a winter I'll not soon forget. For three weeks, day and night, it's been below freezing, often in the single digits. The snow is deep and so hard that when you walk on it you barely break through the crust. The lawn in front of the house is strewn with disks of ice from the dogs' water bowl. After a string of mild winters, here comes the payback, a reminder that the easy days don't roll on forever.

The Farmer

Much of the driveway is slick with ice and impassable without four-wheel or at least front-wheel drive. The section that slopes away in front of the tractor shed is particularly treacherous, especially when a thin coating of fresh snow disguises what's underneath. When I go to the barn to feed the chickens and cats, I strap a pair of Arctic Spurs onto my boots. These are sets of spikes, like crampons, that bite into the ice and greatly reduce the chance of a fall.

In the last few weeks, two of the chickens have succumbed to the cold: first an old hen who had given many years of uncomplaining service, and then Lazarus, the older of the two roosters. Both were found dead on the ground, frozen solid. I threw their carcasses out into the cow pasture where the crows will find them. For Lazarus it is probably just as well that the end has come. Since the coup d'état· in the fall in which he was bloodied and overthrown by the newcomer, Jose, his life had taken a dramatic turn for the worse. But at least he sired an offspring before his demise who may eventually grow up to challenge his nemesis.

Of the three hens who sat on eggs last summer, all presumably impregnated by Lazarus, only one saw her work through to the end. The others became restless and abandoned their clutches a few days before term. And, as if to aggravate this rather shabby performance, Rosie, the one hen who remained on her eggs, managed to successfully raise just one chick. As it grew in size, the young bird blended into the flock and I didn't know what sex it was until about a month ago, when it became clear that Lazarus had fathered a son who looked remarkably like him—a pale, speckled bird with big feet and a handsome comb. Now that his father is gone, it seems fitting that this young fellow be named Son of Lazarus.

The cats, Bernie and Samantha, have grown thick coats and seem to be faring well enough. To stay warm, they curl up together in empty wooden crates or cardboard cartons in the lower barn. I placed a soft pillow inside a carton in a corner of one of the horse stalls and they have taken to it as a favorite spot. When I enter the barn to feed them, they dash toward me and then pace about expectantly, meowing wildly, with their tails high in the air. It is clear that they are hungry, much more so than in warmer weather when they are able to take at least partial responsibility for their own sustenance. Baby rabbits, field mice, voles, and other unwitting small creatures become a significant part of their diet. But, with two feet of snow and freezing conditions, baby rabbits are nowhere to be found.

Aldo and Tiki, the white dogs, are unfazed by the frigid weather. On the coldest nights, they stay outside on the lawn or in the driveway guarding against intruders both real and imagined. Tiki is especially vigilant. She moves around the yard and barks frequently, even during the darkest storms. Who knows what undesirables might be lurking on the edges of the farm? Her bark is a constant warning that those with ill intent should look elsewhere or be prepared for combat.

Most of our neighbors see black bears from time to time, but I have never encountered one on the farm, though we are surrounded by woods and situated a quarter mile from the road. I assume no bear in possession of its faculties would wish to encounter these two fearsome canine guards. I like to imagine their response. At lightning speed, they would come, from different directions, close to the ground, each a hundred-odd pounds of muscle and sinew, snarling and barking, side-stepping and lunging with open jaws and bared teeth. It is their role in the universe. It is what they long to do. To witness such an encounter, which I'm sure would end with the bear's hasty withdrawal, would be a rare privilege.

For Kuri, my first and favorite dog, this winter is particularly cruel. In addition to the bone-chilling cold, he must also suffer the indignities of advanced age. And they are many. In the last few months he has lost much of the feeling in his hind legs, causing them often to collapse under him, leaving his rear end splayed ignominiously on the ground. His hearing, which has been failing for some time, is now so poor that we have to beep the horn repeatedly to avoid running him down on the driveway. His bark is a faint semblance of what it used to be. Some of his teeth are gone and those that remain are not strong enough to crush bones to extract the tasty marrow inside. To these harbingers of decline must be added the irritability that comes with weak muscles, aching joints, and the inability to do many of the things he once took for granted. But perhaps worse than any of the above are issues related to his evacuation processes.

Since Kuri was always discreet in these matters, it pains me to acknowledge that, of late, he has begun to soil his pants, so to speak. Of course he has no pants, which is just as well, though I have on occasion tried to induce him to wear one of my old sweaters to protect against the cold. But even without pants, random and unexpected bowel movements are becoming a real problem for him. When he curls up at night to fall asleep, there's a good chance he'll awaken in the morning lying in or next to a pile of his own waste. This infuriates him, and must be all the more vexing since his keen sense of smell, unlike many of his other faculties, seems not yet to have abandoned him.

I'm sure he believes that one of the other dogs—most likely Aldo, whom he has never really warmed to—is responsible for fouling his nest. With obvious distaste, he struggles to lift himself above it, then stands in place, sometimes for several minutes, during which time he emits a string of hoarse, barely audible barks as a way of expressing his displeasure. The other dogs, if present, keep their distance and quietly watch in a deferential manner. I've wondered sometimes if there might be a hint of compassion in their soulful brown eyes. Kuri usually follows his vocal harangue with attempts to hide the offensive stuff, employing his nose to cover it with clumps of hay or small stones. Eventually he wanders off and looks for another spot to lay down his ancient body.

That leaves me to clean up his mess, which I do on a daily basis. Fortunately his turds are usually firm and well formed and, of late, are frozen solid by the time I get to them, so they're not that difficult to remove with a shovel. But it is an unhappy state of affairs for Kuri and I am sensitive to the indignities he must endure. Yet I see no indication that he is ready to give up his life. He still greets me in the morning with a wagging tail and nudging nose and accompanies me wherever he can. On his better days, if it's not too cold and the snow is not too deep, he expends considerable energy inspecting the nearby hedgerows and field perimeters to determine what creatures might have passed in the night. When I work in the greenhouse he's sure to be there with me, sleeping under one of the benches. As befits an old dog, most of his day is spent in rest.

Were Kuri still in control of his bowels I might open the front door of our house at night and give him the pleasure of a heated room. But then again, I might not. I am conflicted on this matter. For the fourteen years that Kuri has been with us on the farm, he has always lived outside and has weathered, without complaint, what chance and nature have dealt him. Until recently he enjoyed the shelter of a doghouse and the warmth within generated by his own body and sometimes a small heating pad. But his incontinence has made the doghouse unpalatable to him and the evident stiffness in his long legs and spine make it hard for him to get in and out.

A month ago, when the weather turned sharply colder, I took some bales of hay and built a small enclosure in a back corner of the tractor shed. Kuri immediately recognized that my efforts were being made on his behalf and expressed his approval by enthusiastically raising and lowering his head and pulling back his lips to form an expression that I have always interpreted as a grin or smile. As soon as I was finished, he moved in and started to shape the new spot a little more to his liking. Within an hour he had ripped apart one of the bales and spread its contents across the gravel floor, creating a rather appealing circular bed. This is where he now prefers to spend his nights.

When the temperature is below twenty degrees and after he has finished his evening meal, I wrap him in an old woolen army blanket. As I tuck him in, he curls up tightly and doesn't move, but emits deep sighs and grunts which I'm sure indicate approval. Often he goes to sleep completely covered and would appear to a stranger to be no more than a motionless pile in the corner of the shed. Sometimes in the middle of the night, if I shine a flashlight on the pile, I can see the end of a paw or his long nose sticking out for a little air.

During snowstorms, if the wind is blowing from the south or the east, the entire inside of the tractor shed may receive a coating of snow, sometimes several inches deep. I've gone to Kuri in the morning and found him completely buried under a double blanket, one furnished by me, the other by nature. Under it all he can be surprisingly warm and well insulated.

The worst nights for Kuri are the ones without snow, when the temperature is close to zero and a strong wind is gusting through cracks in the shed. On nights like these the air bites into the skin. If I step outside without a woolen hat and heavy gloves, within a minute or two my ears feel like wedges of ice and my fingers ache with the cold. On such nights I make sure that Kuri is well fed and well covered. An old sleeping bag gives him added protection and he is treated to large chunks of meat usually gleaned from road-killed deer or stillborn calves salvaged from Eddy Bennett's dairy farm. I keep a good supply of these rations in the freezer, knowing how much the three dogs appreciate them.

When I give Kuri meat, he devours it ravenously, hardly bothering to chew it at all, as though it were his first meal in a week. At the same time, he shuns his accompanying dry dog food, though he will revisit it later. Kuri also takes a couple of pills: a glucosamine chondroitin smeared with vitamin E for his arthritis and, every second night, a steroid that the vet prescribed for his back legs. While the pills, especially the steroid, appear to have had a salubrious effect, they are not high on Kuri's list of desirable foods. However, when hidden inside large pieces of meat or cubes of lard, he readily ingests them.

This is a winter I'll not soon forget. For three weeks, day and night, it's been below freezing, often in the single digits.

But still, one might ask, since I go out of my way to provide him with meat and wrap him in a blanket, why not allow Kuri inside on these cruelest of nights? I am aware that some friends and neighbors think me a little heartless for not extending him this final favor. After all, unlike Aldo and Tiki, he is a short-haired dog, less well adapted to zero-degree weather. With his sleek, black coat, his long, skinny legs, and his coonhound blood, he is more suited to ranging through southern forests, in hot pursuit of his quarry.

But then I remember—he is a dog, not a human; nature is his true realm, not a heated room in a house. The outside world brims with familiar life: The sounds and smells of other animals, the wind worrying the trees on a dark night, the summer sun that warms the flesh, and the biting cold of winter—all these have shaped his deepest being and his wild spirit. They are his world.

As Kuri approaches the end of his long and rich life, is it not right to let him remain in his native element? In time the earth will take its toll, as it has done with so many dogs before him and countless generations of his wolf brethren. In the primal blood that courses through his veins, there must run the knowledge that, barring some premature misadventure, the end will not be warm and easy in a closed room, but will instead be cold and hard and solitary. These are the terms of the bargain he has made, his fiery compact with life.

The pack must maintain itself and move on. The aged and the weak are no longer needed. Life will renew itself in its own fashion, without remorse or self-pity. Is this not Kuri's destiny and the destiny of every other wild thing, and perhaps, in the end, ourselves even?

Is there not dignity in this knowledge and an almost too blinding beauty—this final letting go of earthly pain, this dissolving of the self into the great roiling river of existence? Is this not the knowledge that teaches us a kind of tenderness, and even love?

ON the EVE of WAR

(February 2003)

Woke up this morning with a gentle snow falling and a quietness on the land. Almost everything is white again, or dark and white. Color is nowhere to be found. The drooping limbs of the big Douglas fir at the corner of the lawn, the lateral branches of the old mulberry tree outside my office window, the empty bird feeder and the cage of suet: All are coated with snow. Even the wrought iron lawn chairs are mostly white, except for their dark legs and backs. Arranged in a semicircle, they seem to await guests at a winter garden party, offering inviting cushions of snow.

The drabness of the old snow, the dark patches of ice and bare gravel on the driveway, the footprints and animal pawings, the scattered droppings and debris that squirrels have taken from the compost pile—all these are gone, as if washed away, leaving a cleaner, more pristine land.

I've cancelled my 9:00 AM dentist appointment in Middletown and will forgo the planned trip to the post office. The mundane business of life will have to wait. This is a day for solitude and reflection. No guests will come to occupy the lawn chairs. No visitors will chance the long, steep driveway to the house. Daunting even at the best of times with its ridges, rocks, washouts, and potholes, when covered with several inches of snow, the driveway is virtually impassable.

I'm not unhappy to be sequestered away in this soft, white world where the only visitors will come silently, on wings, looking for a little suet to put in their tiny bellies.

The world of men is bearing down too heavily of late and I am glad to have some respite from it, if only for a day. All the talk of war is unsettling. The benighted willingness of the people; the lip-smacking bravado of the leaders as they mobilize vast forces to attack a small despot who constitutes no threat to our security. When before has such a crippled individual been so lionized, so imbued with nefarious global designs? The old themes of empire, conquest, and plunder are in the air. The dogs of war are about to be unleashed. There is no balance. With our great wealth and destructive might and our overweening confidence that God is on our side, what dark and terror-filled age might we plunge this small world into?

A MAN and HIS TRACTOR

have just bought a new tractor, or, more precisely, a new old tractor, and am feeling happily "chuffed"—to use a common New Zealand expression that roughly translates into "being quite pleased with yourself." Which indeed I am—there's nothing quite like buying your first tractor, or your second, or your third, as is now the case.

This latest acquisition is a John Deere 4040 manufactured circa 1980. With a 90-horsepower engine and a weight of ten thousand pounds, it is a large beast—considerably bigger than my other two tractors. The rear tires are over five feet high and two feet wide and

The John Deere 4040

probably weigh five hundred pounds each. To get to the driver's seat, you have to climb up three steps.

What drew me to this behemoth is a feature that my other tractors do not have—a "creeper gear" or supplemental transmission that enables it to go extremely slowly—all the way down to a third of a mile per hour. This is slow enough that you can dismount while the tractor is still moving, answer nature's call if need be, and climb back on again without having to walk more than a few paces.

The very slow speeds available through the creeper gear will make it possible for me to start using a water-wheel transplanter—an ingenious implement for vegetable growers that I have long wanted to own and, coincidentally, have recently acquired. To put things in perspective: every year, over the course of the growing season, we insert into the soil, individually, well in excess of 100,000 seedlings, sets, and cloves. Until now, this planting has always been done by hand. It is slow, laborious work that requires serious bending and often crawling on hands and knees. Willing volunteers are hard to come by. With the new tractor and the water-wheel transplanter, the planting phase of my operation, at least for some crops, should be faster and a lot easier on the knees and back.

The transplanter, which will be lifted and pulled by the new tractor, is fitted with gently spiked wheels that make holes in the soil at a designated spacing. Each hole receives a shot of water from two eighty-gallon tanks mounted above the wheels. My crew—up to four persons at a time—will sit comfortably on seats provided just behind the wheels. They will have trays of seedlings or onion sets or garlic cloves, or whatever we wish to plant. All they have to do is drop a plant or clove or set into each hole. The key to success is that the tractor go slow enough that the human planters don't fall behind and find themselves in a situation comparable to the one Lucille Ball experienced on the confectionery packing line, in which she was compelled to find other options than those intended by management for the candies coming her way.

The new tractor will also serve as a good backup for my midsize tractor—a 55-horsepower John Deere 2350. This machine is smaller but has four-wheel drive, which gives it enough power to pull a five-shank chisel plow at a depth of fifteen inches or operate a large rear-mount rototiller—the two most important implements in my tillage system. The 2350 also has a fully functioning hydraulic bucket or front-end loader, which I would be hard-pressed to live without.

Many years ago, when I went shopping for my first tractor, I wasn't sure I wanted to incur the added expense of buying a tractor with a bucket. When I mentioned my uncertainty to a testy old farmer who was offering for sale a small Massey Ferguson tractor close to himself in age, with a severely abused bucket, he informed me in a rather dismissive tone that a bucket was more useful than a wife. Fortunately, my wife, who was with me that day, was out of earshot at the time. While I certainly disapproved of the old farmer's sexist remark (and today, having time and experience on my side, still in no way agree with him), he did impress upon me the desirability of a bucket. They are extraordinarily useful, and once you've had one, there's no going back.

Whether on principle or for more practical reasons, I did not buy the tractor from that old chauvinist, but soon afterward I acquired a 1960s vintage Allis Chalmers with a primitive bucket—it can be lifted hydraulically but relies on gravity to empty itself. I still own this tractor and use it frequently for lighter, less demanding jobs. It rarely fails me and, when it does, it is quite amenable to repair.

These days I rely heavily on the bucket on the John Deere 2350. I use it to pick up and transport heavy objects or material such as gravel, cement, or sand. I use it to turn compost, push snow out of the driveway, and move rocks out of our fields. I use it to push over dead trees or uproot living ones. I've even used the 2350 bucket to demolish a fairly large shed and position a walk-in cooler exactly where I wanted it.

Before becoming a farmer, I never spent much time thinking about tractors, nor was I especially interested in owning one. I knew they were good for pulling heavy objects like plows at slow speeds, but beyond that I accorded them limited utility. Today I view tractors very differently. I have come to regard them as almost indispensable in my daily life, perhaps even more so than buckets. And of course, tractors confer on their owners a certain prestige and enhanced sense of power that is a real bonus as age and time exact their cumulative toll.

Generally, I am not a man who cares much for machines. I can scarcely differentiate one model or make of car from another. And most electronic gadgetry leaves me cold. But tractors are something else, especially old ones. They are tremendously sturdy and have a functional, workmanlike simplicity about them. Keep a tractor supplied with fuel and oil, change its filters once a year, and it'll do the job for you, day in and day out, in every kind of weather. The newer tractors with air-conditioned cabs, GPS, computerized electronics, and multiple options are not for me. I go for the oldies, without frills, where what you see is what you get.

I am reminded of a poem by the New Zealand poet R. A. K. Mason that I read more than forty years ago in a course on New

Zealand literature and probably haven't looked at since. For some reason, it has stuck with me.

Prelude
This short straight sword
I got in Rome
when Gaul's new lord
came tramping home:

It did that grim
old rake to a T—
if it did him,
well, it does me.

Leave the thing of pearls
with silken tassels
to priests and girls
and currish vassals:

Here's no fine cluster
on the hilt, this drab
blade lacks lustre—
but it can stab.

—R. A. K. MASON (1905–1971)

I assume Mason was alluding more to his preference for a certain blunt style of language than a tool of physical violence. In any event, his tight little poem, with not a single superfluous word, nicely captures some aspect of the feelings I have about my old tractors.

A beginning farmer soon learns that a tractor by itself is not enough. It is, in fact, little more than a powerful, slow-moving, single-occupancy vehicle with disproportionately large rear wheels. Only when coupled with the various complementary implements designed for it does a tractor's great utility emerge.

Whether on principle or for more practical reasons, I did not buy the tractor from that old chauvinist.

Plows, disks, planters, mowers, cultivators, rototillers, manure spreaders, and corn choppers are just a few of the many tractor implements available to a farmer. A tractor without implements will not get you far, and implements without a tractor are virtually useless.

Most tractors made in the last fifty years are designed to perform four important functions. The first and simplest of these is straightforward pulling. Tractors will pull just about anything you want to pull—be it a wagon, a cart, another tractor, or a drag implement such as a disk or spring-tooth harrow. You simply attach the object to be pulled to the tractor's tow bar using a heavy pin, and off you go.

A second and slightly more complicated tractor function is dependent on an invaluable feature known as the three-point hitch. This operates through two sturdy swinging arms in the rear of the tractor. These arms can be raised and lowered hydraulically. A third connection is made above the hydraulic arms using a separate telescoping, adjustable arm known as a top link. The three-point hitch feature enables a farmer to pick up heavy implements such as plows or rototillers, transport them to the field, and then lower them into the soil where they can be put to use. By raising such implements off the ground you avoid tilling or plowing your driveway en route to the field—always preferable.

There is tremendous lifting power in a tractor's three-point hitch hydraulic arms, but, naturally, larger and more powerful tractors are able to pick up more weight than smaller ones. My new water-wheel transplanter with four people seated on it and two eighty-gallon tanks full of water will probably weigh around three thousand pounds. The John Deere 4040 should have no trouble lifting this amount of weight, whereas my 38-horsepower Allis Chalmers wouldn't stand a chance—the front wheels would lift off the ground long before the transplanter did. The John Deere 2350 might be able to pick up this load, but it's a moot point since this tractor doesn't have a creeper gear and therefore cannot go slowly enough for transplanting. Also, it exhausts out the back, as opposed to straight up like most tractors. At some point, the human transplanters would surely be overcome by diesel fumes and quite likely fall off their seats.

A third important tractor capability involves the use of a hydraulic cylinder and hoses to raise and lower certain drag implements that cannot be attached to the arms of the three-point hitch but usually have wheels for transporting them. A disc is a common example of this type of implement. Hydraulic hose lines can also perform such tasks as opening and closing the rear gate on a manure spreader or adjusting the position of a planter on a steep slope.

The fourth essential feature on a tractor is the power take-off shaft, commonly referred to as the PTO. This grooved shaft, about 1.5 inches in diameter, protrudes from the rear of the tractor and provides power to implements with moving parts, like rototillers and brushhogs (a brushhog is a large, single-blade mower capable of cutting down everything in its path, from tall grass to saplings an inch or more thick).

When the PTO shaft is engaged, it turns with relentless power. It is like having a second engine to operate an entire class of implements. In addition to tillers and mowers, the PTO shaft can be used to power tree shredders, log splitters, and post hole diggers. PTO shafts can even run specially equipped pumps and generators. But beware: PTO shafts can be hazardous to your health and should be treated with great respect. If you get too close to a turning shaft and it catches your shirtsleeve or jacket, it's quite likely your arm will be separated from the rest of your body at the shoulder socket. There are more than a few one-armed farmers who leaned over too far to adjust an implement without first disengaging their tractor's PTO shaft.

Many implements depend on more than one of the above features common to most tractors. A rototiller, for example, is lifted by the arms of the three-point hitch and its moving tines are powered by the PTO shaft. A disk is usually pulled by the tractor's tow bar and raised and lowered by the hydraulic cylinder and hoses.

The world of tractors and tractor implements is vast and not well understood in a single season of farming. I often tell my workers that I can probably teach them how to drive a tractor in ten or fifteen minutes but that it may take several years before they learn how to use one to good effect in its various capacities.

Some might think my attachment to and dependence on tractors a little unbecoming for a small organic farmer. Surely it would be more fitting and environmentally correct to embrace age-old agrarian hand tools and avoid these large, fossil fuel–burning machines. Or else rely on horses or oxen and associated tillage equipment to work the fields. I will admit to some small guilt over this latest tractor purchase. Perhaps it is a little excessive to use three tractors to farm just twelve or fifteen acres, even if each *is* suited to its own specific tasks.

When I mentioned the new tractor to a holier-than-thou friend of mine, rather than congratulate me or share in my enthusiasm, he right away remarked that my acquisition would result in more pollutants entering the atmosphere. After recoiling from this unexpected blow, I asked him how he felt about his small family's three cars, riding mower, and many electrical appliances that presumably derive their juice from oil- or coal-burning power plants somewhere down the line. Rather than engage me further on this line of inquiry, he abruptly changed the subject. But the damage was done, and since I was deprived of the opportunity to defend myself, I am obliged to respond in public to this private assault.

Our place is not every vegetable grower's dream farm. Before we came along, it knew only cows, and in the previous twenty or thirty years had earned a scant living for the few families who worked it. The young dairy farmer who sold us the place milked his cows in the morning and evening and drove a beer truck during the day (and incidentally had three tractors to help him). Our farm has plenty of character and lots of rough, wild, untillable land. While it suits me well enough, it lacks the fine, alluvial bottomland that vegetable growers prefer. Much of our soil is on the heavy side and peppered with rocks. Many of the fields have a fair degree of slope to them. Nearly all are surrounded by woods, which often deposit dead trees around their edges. To take on this land with just hand tools would be a very daunting task. A dozen men with shovels and picks might put in a week of hard labor to accomplish what I can do with a tractor and rototiller in a couple of hours. If you could find people to do this work, which is highly unlikely, you couldn't afford to pay them, given the economics of farming in these modern times.

Each of my three tractors has its own specific role to play. With the exceptions of transplanting and bucket work, the two larger John Deere tractors can perform many of the same functions. One therefore plays the important role of standby or backup in case the other suffers a temporary breakdown. The various tasks assigned to each tractor are as follows:

Allis Chalmers D15 (38 horsepower):
    Cultivating large-area plantings for weed control
    Mowing with brushhog around edges of fields
    Cutting furrows for planting
    Laying black plastic
    Loosening soil around garlic before harvest
    Hilling and harvesting potatoes
    Spreading compost

John Deere 2350 (4-wheel drive: 55 horsepower):
    Chisel plowing to incorporate compost and cover crops, control weeds, and generally loosen and aerate soil
    Rototilling for seed bed preparation
    Disking to control weeds and create level planting areas
    Snowplowing in winter
    Multiple uses of bucket

John Deere 4040 (with creeper gear; 90 horsepower):
    Transplanting with water-wheel transplanter
    Backup to John Deere 2350
    Restoration of self-esteem when confronted with the limitations of an aging body and other insults of time

Though the three tractors and their implements are integral to our farming operation, they definitely do not dominate our daily work—most of which is, in fact, done by hand. The tractors are used primarily for field preparation, maintenance work (such as mowing along the edges of fields), and some large-scale planting and cultivation. Normally, I do almost all the tractor work unless I have an experienced operator on payroll. In my typical sixty-five-hour workweek during the growing season, I might spend six to eight hours on a tractor. The rest of the time I'm on foot with everybody else.

I like my tractors and I need them. They help me grow food and stay on the land.

Fuel consumption is not excessive; 150 gallons of diesel will usually last me for a year, though this will probably go up with the John Deere 4040 on the scene. The Allis Chalmers runs on regular gasoline and consumes about 100 gallons a year.

One final point: a good tractor is a good investment. Especially if it's a John Deere, which is generally regarded as the Cadillac in its class. Ten years ago, I paid $10,500 for my John Deere 2350 (it was about twelve years old at the time). Today, I could probably sell it for $11,000 or 12,000 without much trouble.

But put aside these self-serving justifications. In the end, I like my tractors and I need them. They help me grow food and stay on the land. They never insult me or question my motives and they seldom let me down. I'll openly admit to a late-life romance with them.

THE HIGH PRICE of MILK

On the afternoon of January 26, 2004, Eddy Bennett and Kevin Miedema were hauling felled trees from a wooded patch on Eddy's farm. It was a cold, clear day with a few inches of hard-packed snow on the ground. The two men were using a tractor and chains to drag trees out of the woods and across a couple of fields to a convenient spot by the road where they would be sawed up and split for cordwood. It was a way to make a few extra dollars to supplement a lean milk check.

They were working on the last tree of the day—a large white ash— before they would return to the barn to give the cows their evening rations. Kevin was on the tractor and Eddy was walking behind with a chainsaw in one hand, keeping an eye on the tree to see that it didn't slip loose from the chains. It was getting dark and the ground was uneven. Some obstruction—perhaps a hummock or a boulder—under the frozen snow caused the tree to sharply buck over to one side. A heavy limb struck Eddy in the head. Kevin felt the sudden jerking movement of the tree and looked back to see Eddy propelled through the air from the force of the blow. He stopped the tractor and ran to his friend lying in the snow. When he saw Eddy was unconscious, he dashed back to the barn and dialed 911 and his parents' house, just a mile away.

Kevin's father, Jim, immediately rushed down with sleeping bags. In the frigid air the two could see that Eddy's head was swelling. They covered him with the sleeping bags and waited—the father with Eddy and the son out by the road to direct the emergency crew to the scene. When the ambulance arrived, the EMTs quickly determined that Eddy's injury was too severe for nearby hospitals and summoned a helicopter. They put Eddy on a stretcher and tried to give him oxygen. Though his eyes were glazed and he appeared unconscious, he resisted the oxygen mask.

The helicopter took Eddy to the Westchester Medical Center in Valhalla, where he was admitted to the trauma unit. For a week he lay on life support and there was much hope that he would survive the massive head injury he had sustained. But that was not to be. On the eighth day Eddy died. All of us who knew him were shaken to the core.

As we advance through life, most of us accrue a deeper understanding of mortality. No matter how sunny our disposition, how privileged our circumstances in the world, it becomes increasingly apparent that loved ones, family and friends, those we are closest to, do not stay forever. As time passes, we experience loss with greater frequency. It is one of the drawbacks of living a long life. Most of us learn to bear this bitter knowledge and tuck it away, if we can, in a far corner of the heart. We adapt; we accept; we get on with our lives. Perhaps we become wiser in the ways of the

It's a Long Road
to a Tomato
216

world and the impermanence of things. Still, death has a way of taking us off guard. With supreme randomness, it strikes most unexpectedly, even those in the prime of their lives.

I first met Eddy Bennett in the fall of 1986, a couple of months after we bought the farm. Back then I was still living five days a week in New York City, working at an office job in White Plains, and coming up to the farm on weekends. It was probably a Saturday afternoon in November. Eddy was over at our place and I remember standing with him in the lower barn with the milking stanchions around us. After exchanging a few pleasantries we worked out an arrangement for him to rent some of our fields to grow corn for his cows. Since I was expecting to plant just one acre of vegetables the next year, there was plenty of land available. He proposed a price of $25 per acre per season and I agreed, having no idea of the prevailing rate for rental farmland.

The subject then shifted to manure, which I was eager to obtain. He offered to deliver and spread cow manure for $10 a load. That seemed like a very good deal, especially when he told me that each spreader load would weigh three or four tons. In the eighteen years that I knew Eddy we never changed the terms verbally agreed to on that day. Each spring we settled up in his parents' kitchen and a few dollars changed hands. It always seemed more like barter than business.

But back to that Saturday afternoon in 1986: before Eddy left, he casually asked me what we planned to do about our water. Being unaware of any action I needed to take with respect to water, but already feeling slight apprehension, I asked him just what he meant by that question. He tapped the pressurized water tank that he just happened to be leaning against and said: "This is going to freeze without cows in the barn." That didn't sound good, but what did cows have to do with the water tank freezing? Without alluding in any way to my city-boy ignorance, he gently explained that the body heat generated by a barn full of cows would keep the temperature of the water in the tank well above freezing. Now that there were no cows (they were sold at auction before we moved in),

In the eighteen years that I knew Eddy we never changed the terms verbally agreed to on that day.

there was no source of heat in a very-much-less-than-airtight barn. The inevitable result would be frozen water pipes and a frozen tank. It was just a matter of time. I was beginning to catch on.

Looking at the pipes feeding into and out of the tank, Eddy quickly deduced that water came from the farm well directly into the pressurized holding tank in the barn. One pipe distributed water throughout the barn for the cows to drink and for general cleaning purposes. Another line went underground to the house. In other words, all the water from the well went through that holding tank in the lower barn. When it froze, which it most definitely would, and perhaps quite soon (we had already experienced several frosts), we would have no water in the house or anywhere else and the pipes in the barn would freeze and crack. During my apartment-dwelling days in New York City, this type of knowledge and application of logic was not required. When problems arose, one simply called the superintendent or landlord. In the friendliest way, Eddy was beginning my education in country living.

Within an anxiety-ridden week, my wife and I had assembled a small team (a local backhoe operator and two plumbers) to remedy our problem. Eighteen hundred dollars later, a new water tank was installed in the basement of the house and the water rerouted directly from the well to the new tank, then back to the barn, with a shutoff valve installed so we could drain the barn pipes each winter. It all happened just in the nick of time. On the last afternoon, as the backhoe operator was backfilling the four-foot ditch in which the new pipes were buried, the soil was already freezing into large cement-like clods. We were watching with some measure of satisfaction and relief as Eddy showed up to take a look. I vaguely remember that he smiled with approval.

As our friendship developed over the years, that same open-hearted spirit never left it. Many were the times I called Eddy to ask for help or advice. He always gave both willingly and never made me feel like the ignorant greenhorn I was. In the early years, I got my tractor and disk stuck in a wet spot in one of the back fields. I walked to the barn, found a couple of jacks, a shovel, several boards, and some rope, and drove back in my pickup, hoping to dig myself out. Before long the pickup was stuck, too. After a couple of hours of digging and jacking, sweating and cursing, I gave up in a state of intense frustration. I returned to the house, swallowed my pride, and called Eddy. A half hour later he showed up on his big four-wheel-drive tractor with a broad grin on his face. Within twenty minutes he had pulled my tractor, disk, and truck onto dry ground and was heading back to milk his cows. It was just a neighborly thing to do.

A few years later, when my tractor rolled over on a slope, almost crushing me and putting an end to my farming career, he came right over with chains and pulled it back to its upright position. Before leaving he warned me about the danger of riding on slopes with the front-end loader lifted too high and full of rocks.

Eddy Bennett was the kind of guy who didn't change with the times. There was a constancy about him, like the contours of the land he farmed or the course of the creek that ran under the bridge below his house. He always referred to his land simply as "home" and seemed satisfied with his lot. It was enough for him. There was something about Eddy that led me to believe he would always be here as a farmer, a neighbor, and a friend who would willingly lend a hand when I needed it most.

"Hi, neighbor," he would say, with a tilt of his head and a warm grin, whenever I stopped by his barn to negotiate the delivery of a load of cow manure or to pick up a bag of used baling twine he had saved for me. Regardless of how busy he was, there was always time for a chat. He'd ask after my crops or my crew—amazed that young college-educated men and women would come all the way from West Virginia and Iowa and California to labor in the fields of a small vegetable farm in his hometown.

To be sure, Eddy and I had very different backgrounds, very different worldly experiences: I had traveled over large portions of the world and lived seventeen years in Manhattan; he had never strayed far from home and not once visited the great city just sixty-five miles to his south and east. I have seven years of college and

There was a constancy about him, like the contours of the land he farmed.

two graduate degrees; Eddy finished high school and took some courses at the local community college. I am a married man, more than once, and have had my share of girlfriends; Eddy was single and shy in the company of the opposite sex—he had no female companions that I knew of. I am an organic grower, frequently at odds with the agricultural establishment and quick to express my left-leaning views; Eddy grew up using chemicals and did not seriously question the long-term implications of their use. His main social activity was going to church on Sunday, though he rarely spoke about this.

There were real differences between us, but we were both farmers in a town with very few farmers left. That seemed to be a sound enough basis for friendship. If anything, our differences made each more interesting to the other. There was much I could learn from Eddy and he was curious to hear about my vegetables and what it was like selling them on the streets of New York City. He remembered what I told him and often asked follow-up questions weeks or months later. There was always plenty to talk about and always room for a laugh. What stood out most about Eddy was his gentleness and his willingness to accept people if they were straightforward and honest with him. It was as though some youthful innocence had never left him.

During the week before he died and for a week or two afterward, before arrangements could be made to auction them off, Eddy's cows had to be milked and fed and the barn kept clean. It was more than Kevin Miedema could do alone, so a group of neighbors—all former diary farmers—pitched in. They took turns. Some got up at 3:00 AM to do the morning milking before going off to their day jobs. Others came in the evening, after work. All went about the task quietly, reflectively, absorbing the loss of a friend and, perhaps more than that, a way of life.

Though most of the half-dozen mostly middle-aged men who came to help had grown up on dairy farms and knew the rhythms of farm life, all had moved on to other work. Milking cows could no longer bring in enough money to support them and their families. As they came to the aid of a dying friend, they also returned to their younger days and, for those born into it, the strangely comforting sounds, smells, and feelings of a barn full of cows chewing hay and giving milk. Probably for most of these men it would be the last time.

The loss of Eddy Bennett has moved this small former dairy town one step closer to what appears to be its inevitable exurban destiny. Soon after his death, Eddy's nephew, Rick Hensley, moved into the old farmstead house. As a youngster he had often visited his uncle and worked at his side. They enjoyed each other's company. Rick has told me he doesn't want to see the farm, which has played such a central role in his family's history for four generations, go to development, but the responsibility and cost of maintaining it is large and Rick is not a farmer. He has his own landscaping and snowplowing business to take care of. For now the farm is intact and looks pretty much like it did when Eddy was alive, but the cows are gone and are not likely to return.

Soon after his death, one local woman said to me, "I hate to use words like this, but Eddy Bennett was as close to being an angel as anyone I've ever known." It sounds a bit corny, ridiculous really, and would certainly have brought out a great guffaw from Eddy and an emphatic shake of his head, but it captures some essential quality that those of us who knew him will long remember.

1958-2004

WORKING MAN'S MESCLUN

Twenty years ago, "salad" was a limited concept. Even in a good restaurant, if you ordered a salad with your main course, you'd likely have received a plate of chilled iceberg lettuce of unknown vintage, with a few wedges of hard tasteless tomato and perhaps a slice or two of red onion to go with it. Flavor, if the word can be used in this context, was attained through one of several exotically named but often unnaturally overpowering dressings. Croutons, shredded cheese, and artificial bacon bits may have been offered as extras. The experience was not uplifting.

Though virtually unknown in this country in the early 1980s, today the mixture of washed baby salad greens known as mesclun is common fare. (The word "mesclun" comes from Provence, France, where it refers to a seasonal mix of tender young lettuces and other edible greens, both cultivated and wild.)

I first encountered the American version of mesclun in my early days at the Union Square farmers' market. I was selling heads of lettuce for a dollar each. Across the aisle from me a farmer with a booming voice and a red kerchief around his neck was dealing out a mysterious salad for the amazing price of $12 per pound. And he had no shortage of customers.

For a time I watched my neighbor with a mixture of envy and awe. Then I began to wonder if I, too, might participate in this gastronomic phenomenon. By the end of my second season the lure of $12-a-pound salad was too much to resist. I was ready to jump in.

That winter I did a little research, studied the seed catalogs, and selected the ingredients for my first mesclun. As it turned out, I was already growing most of them anyway. I just hadn't got around to combining them in a salad mix.

I soon learned that there is no single formula for mesclun. Whatever fresh greens are seasonally available, look good, taste good (at least in combination with other greens), and lend themselves to being eaten raw may be included in the mix. Beyond these basic criteria, there are a few more subtle factors to consider. A good mesclun will be visually attractive as well as pleasing to the palate. The shape, size, and texture of the leaves selected, the mingling of colors and flavors—all contribute to the excellence (or otherwise) of the final product on the dinner table.

Subjected to the rigors of the marketplace, over time our mesclun has undergone something akin to an evolutionary process, with numerous small changes and a few wrong turns and failed experiments along the way. (For example, we no longer include shungiku, an aromatic chrysanthemum with a painfully bitter taste, or edible flowers, such as nasturtiums, which add an attractive floral dimension but compromise the storage life of the mix.)

Today, our typical autumn mix includes the following seven categories of greens, always allowing some leeway for fluctuations in availability, quality, and flavor (the proportions are approximate):

Lettuces (from mildly sweet to mildly bitter)—35% of mix
 Examples: Red Oak, Winter Density Romaine, Bronze Arrowhead, Lolla Rosa

Spicy and Mustard-Flavored Greens—20% of mix
 Examples: Arugula, Red Mustard, Green Mustard

Arugula
& Red Mustard

Asian Greens—20% of mix
 Examples: Japanese Mizuna (a very mild mustard), Chinese Tatsoi

Tatsoi & Mizuna

It's a Long Road
to a Tomato
224

Assorted Cabbage-Family Greens—10% of mix
  Examples: Red Russian Kale, Lacinato Kale, Collards

Bitter Greens—5% of mix
  Examples: Curly Endive, Chicory Frisée, Italian Dandelion

Beet-Family Greens—5% of mix
  Examples: Red Chard, Green Chard, Beet Tops, Perpetual Spinach

Sour and Lemon-Flavored Greens—5% of mix
  Example: Sorrel

Other ingredients that, on occasion, find their way into our mix include purslane, lamb's-quarters, curly parsley, and escarole.

While in most instances small is preferable in the world of mesclun, it is my belief that one should not pursue this goal to the level of an obsession. Leaves that are too small can lack the robust flavor of their larger brethren. We look for leaves that are young and tender and not too big, especially when they belong to plants with strong or distinctive flavors, such as dandelion, kale, and the sharp mustards. With lettuces and a few other mild greens we will accept somewhat larger leaves (and so, it seems, will our customers). Good flavor is what we are all looking for. Because of the slightly larger leaf size in our mix and the emphasis on flavor over "designer sleek," we've sometimes dubbed our mix "the working man's mesclun."

Not long after I jumped into the mesclun business, corporate farms in California noticed a similar marketing opportunity. Now there's hardly a supermarket in the country that doesn't offer mesclun for $8.99 a pound or some such lowly figure. Never mind that it is an agro-industrial product, presumably grown and harvested in a highly mechanized fashion with little regard for soil health and sustainability. And never mind that it is grossly overpackaged and usually wilted, aged, and sometimes brown around the edges by the time it gets to your dinner table.

The public's expectations have been lowered, and the public's perception is that locally grown mesclun is a bit pricey. For this reason we sell our mix today for $4.00 a quarter pound—only $1.00 more than we sold it for in the 1980s. Our profit margin has suffered a little, but this decline has been offset by less modest price increases on almost all other items at the stand. Our intent has merely been to keep pace with inflation and selflessly further the cause of free enterprise.

Though not by temperament a great eater of salads, I do consume my share of mesclun. When dining alone, rather than cook a meal, I'll often reach for a bag of mesclun that made its way back from market to our refrigerator. To beef it up a little, I'll throw in plenty of feta cheese, a half-can of chickpeas, a mild onion or shallot, a thinly sliced clove of garlic, a chopped-up, cooked potato (if one is available), and a can of tuna or Alaskan pink salmon. Smothered with olive oil and balsamic vinegar or perhaps one of Paul Newman's dressings, it's a hearty three-course meal in one.

Salad greens, like other vegetables, are at their very best when eaten straight out of the field. For most people this is not an option. But happily, when harvested, washed, and chilled in a timely manner, mesclun will maintain a high degree of freshness and flavor for several days.

We make our mix the day before we sell it. The process begins midmorning in the field, where we cut predetermined quantities of the various greens into plastic tubs. The tubs are quickly moved out of the sun and taken to a wash room in the barn. (Known as the milk room, this cool, concrete-block room formerly housed a bulk tank wherein the resident dairy farmer stored the day's supply of milk. We've refitted the room with sinks, shelves, and good lighting to meet our mesclun-processing needs.)

By early afternoon, when the sun is overhead, three or four people begin washing the greens in cool well water. This work is relaxing and offers a respite from the midday heat and an opportunity to engage in light conversation or listen to music. During the washing process a fair amount of culling takes place—damaged, yellowed, and oversize leaves must be removed, along with any weeds, insects, or other foreign objects that might have found their way into the tubs. Those greens that survive the washing and screening process are then placed in a large, electrically operated salad spinner, called the Greens Machine. The leaves that don't pass muster are thrown to our flock of chickens, who devour them with much clucking and gallinaceous enthusiasm.

After their short ride in the salad spinner, during which excess moisture is removed by centrifugal force, the greens are placed in clean laundry baskets where they sit and air-dry for another hour or two before being mixed and packed in coolers for market. The whole process takes from three to five hours, depending on the number of people involved and the amount of mesclun we wish to make.

We plant our mesclun in a swath five feet wide by fifty to a hundred feet long. In summer, to stay ahead of the weeds, we transplant thousands of small seedlings into the field. This is a time-consuming process. In the fall, we seed mesclun directly into freshly tilled ground, which goes a lot faster. To keep the supply flowing all season, we must plant most of the ingredients every ten days or so, from April till September.

There are no guarantees. Some plantings come in better than others, depending on temperature, rainfall, and soil conditions. Some plantings are compromised by weeds. Some quickly go to seed in the midsummer heat, causing the leaves that remain to lose succulence and become bitter. The mustard-flavored greens are often attacked by small, shiny black insects known as flea beetles.

For us, fall is the best time to produce mesclun and August is the cruelest month. In the summer heat the quality of our mix may suffer and the supply dwindle or even vanish altogether.

August and, to a slightly lesser extent, July are challenging in many other ways. Important crops such as garlic, onions, and shallots are ready for large-scale harvest and curing. Our worst weeds—lamb's-quarters, pigweed, quickweed, purslane—are making their most determined assaults on our fields. The greenhouse is filled to capacity with cool-weather vegetables, including

Twenty years ago, "salad" was a limited concept.

broccoli, kale, collards, chard, and lettuces, all waiting to be planted in the field. Two picking days each week and two market days in New York City demand much attention. And, of course, the heat and humidity slow us down, sapping our energy and weakening our resolve. Sometimes we simply do not want to look at another weedy, seedy patch of mesclun, nor contemplate the prolonged, nitpicking attention it requires.

Then along comes September. The alliums are harvested and safely in the barn. All the fall crops should be in the field. The days are shorter, the nights cooler. The weeds no longer pose such a threat. Many of the obstacles we faced in the summer are behind us. Our spirits perk up. We are reinvigorated and ready to focus our efforts on mesclun once again.

Farming is so much about seasons and cycles and changes. We feel them in our bodies; we sense them in our moods, our small pleasures, our quiet sorrows. Every day we bring a certain store of energy into the fields. It is not always positive energy, but it flows through us just the same. Each season brings a different set of challenges and possibilities. Only nature remains constant, though we continue to wrestle with her and ourselves. The making of mesclun is just one part of a larger undertaking. It contributes, in its own way, to the diversity and healthful functioning of our small farm.

In the morning, before the workday begins, I like to look out over the fields to assess the progress of the crops and enjoy the rich patchwork of colors they express: the deep burgundy of the mustards and Russian kale, the pale green of the curly endive, the lettuces of varying shapes and colors, the blue-hued green of dandelion—all glistening with dew in the early light. At dusk, under the glow of the setting sun, the scene changes; the colors become richer, softer, more golden. Perhaps a nighthawk crosses the sky or a few geese pass overhead. I step back, breathe in the cool air, and observe the richness of nature's palette displayed before me and, for a time, I am persuaded that all is well on this good green earth.

It's a Long Road
to a Tomato
228

TINY TIM and HIS BOVINE HAREM

Tiny Tim weighs about one ton and is a Hereford bull with all his hardware intact. He is an avid grass eater with a thick neck and a friendly disposition. His head resembles, in both size and shape, a large coal bucket, and his scrotal sack brings to mind a couple of duffel bags that have been spliced together. If I approach him slowly and with reassuring words, Tiny Tim will eat right out of my hand or let me stroke his substantial head. More than once I have been delighted to feel his long, rasplike tongue wrap around my fingers as he gladly receives my offering of a clump of grass or palatable weeds.

Tiny Tim

Tiny Tim was born on this farm five years ago and still frequently spends his summers here. After a year's absence, to my pleasant surprise, he showed up a month ago with a small party of female companions. Without hesitation, as though quite aware that they had returned to familiar home ground, this bovine community took up residence in our pasture and immediately set to work grazing down the green grass of spring. After a winter of eating mostly dry hay, it must have tasted good.

Tiny Tim and the half-dozen Hereford cows and couple of young heifers who accompany him belong to a friend and neighbor, Frank Sgroi, who is a builder by trade but a farmer at heart. Frank, his wife Sonya, and their daughters live on some twenty acres of mostly wooded land about two miles from us. In their very large backyard, Frank usually has eight or ten cows, a small flock of sheep, a few dozen chickens, some rabbits, and assorted other domestic animals. He has told me he would like to get a buffalo someday, and perhaps a miniature donkey for his wife.

Frank takes considerable pride in his livestock. He accords them a fair degree of respect, some affection, and very little sentimentality. In his spare time he breeds them, or rather allows them to breed. He sells some of the resulting offspring, barters others, and gives the occasional one away. He also enjoys eating them, as do we.

The Hereford is an English breed of beef cattle. They are solid, well-proportioned animals with mostly reddish brown coats and clean white faces. Frank favors Herefords because they are docile, well-mannered cows, seldom inclined to aggression. If you're going to keep in your backyard animals that weigh anywhere between one thousand and two thousand pounds, hyperactivity and violent behavior are not traits you'll be looking for.

Frank's cows come into heat about forty days after giving birth, and Tiny Tim is usually on hand to oblige. The calves are weaned at anywhere from five months to a year. When they are about a year and a half old and weigh around half a ton, Frank sells most of his steers for beef. The going rate is somewhere between $1.00 and $1.20 per pound for animals on the hoof. The buyer will likely spend another $600 to have the animal butchered. Tiny Tim is one of the lucky ones.

For most of its history, our farm was a dairy farm. Some of its acreage was used for growing corn and hay and some for pasturing cows. The corn and hay kept the cows nourished in winter while the pasture took care of their needs in spring and summer. The pasture—about twenty-five acres of rough, uneven land—has a creek running through it. It has old stone walls, rusted barbed wire fences, an abundance of large rocks, plenty of good shade trees, several shrubby thickets, and a few wet spots. Its soil is heavy, not well drained, and not well suited to growing vegetables. The highest and best use of this land, from a farmer's point of view, is definitely the pasturing of grass-eating animals.

More than once I have been delighted to feel his long, rasplike tongue wrap around my fingers.

Over the years we've rented our pasture—for nominal sums (usually $100 per year for the entire twenty-five acres)—to several local farmers, most of whom are no longer farming. We've had Holstein heifers, Hereford beef cows, and even horses. All have been a pleasure to host. They've kept the pasture in good shape and helped stem the encroachment of brambles, wild rose, and ultimately forest. They have contributed copious amounts of almost entirely organic manure and added a nice touch of rural ambience. They also offer therapeutic benefits. If I find myself in a disgruntled or tense state of mind, a few minutes alone observing the cows will usually put me back on a more even keel and remind me that my concerns are largely inconsequential in the broader span of things.

◆

I'VE ALWAYS BEEN fond of animals, both domestic and wild, and my fondness for them has only increased with time. Compared to my own species, they have much to admire. They are seldom vain, or greedy beyond their immediate needs, or wantonly cruel. They live within their means and they seem at home on the earth and enviably content with their lot. Though often competitive within their own ranks, one species of animal is rarely given to conquest and dominance over another. Nor do animals burden themselves with future fears or excessive anxieties about tomorrow. As far as I can tell, unlike most of my own kind, no individual or species of animal presumes to place itself above and beyond the rest of creation. That we have gained such preeminence over them seems to me unfortunate.

It's ironic that as a young man, I took considerable pleasure in hunting and killing animals. Did my passage into manhood in New Zealand demand that, or was it some domineering, predatory instinct passed down from distant forebears? In those youthful days there was a strange exhilarated lust in killing and then, later, the beginnings of remorse. I recall the first rabbit I ever shot on my uncle Rodger's farm. I was ten or eleven years old. An older boy, Duncan Cameron, a neighboring farmer's son, whom I looked up to, took me hunting with him at the end of a long summer's day. We were walking quietly in a beautiful river valley—the sun low in the sky, the river slow and meandering with shallow ripples and clear pools, and grassy flats with native manuka trees on either side.

A small rabbit appeared in front of us. My companion looked at me, perhaps to gauge my readiness, and then handed me his rifle. With some hesitation, as though at a crossroads and unsure of which way to go, I pressed the rifle to my shoulder, took aim, and fired. Unharmed, the rabbit hopped to one side and paused, its startled ears alerted to danger but apparently unable to determine from what quarter the sharp, foreign sound had come. I worked another round into the chamber and went down on one knee with a strange, newfound sureness. This time my aim was true. I remember the crack of the rifle and the rabbit's body rising up into the air in a wild fling—its last mortal gesture—before falling back to earth. In that moment I knew the thrill and power and sorrow of the hunt.

Now, perhaps three-quarters of the way through my life, I have largely given up hunting and forsaken the desire to kill. But sometimes, in my daily endeavors to bring vegetables to market, it becomes necessary to exercise that old skill. On such occasions, I try to go about the work with cold-hearted proficiency, a degree of restraint, and perhaps a dash of compassion. More often, though, I take pleasure in observing animals from a distance, without interference, watching them go about their affairs, even if their interests may at times compete with my own.

◆

UPON ARRIVAL AT our farm this year, all of Frank's cows were presumed pregnant. Since then, two of them—Jackie, a thirteen-year-old, and Blossom, who is much younger—have given birth, unassisted, to healthy female calves. The two mothers are protective and seem content in their maternal roles. They know the whereabouts of their calves at all times and move in closely when a stranger approaches. The calves remain shy and unsure of themselves but are rapidly gaining in size as they drink generously from

I remember the crack of the rifle and the rabbit's body rising up into the air in a wild fling.

their mothers' teats. Yesterday I noticed the two calves standing close together, apart from the rest of the herd, looking intently at each other, as though discerning both their genetic and generational bonds.

When Tiny Tim was born, he was accompanied by a twin sister, but she was stillborn. Frank found the three of them together on a Sunday morning. The mother was standing over the dead calf and Tiny Tim was curled up in the grass nearby. Frank attempted to take the dead calf away but the mother would not let him. Flavia and I stood watching; then Kuri showed up on the scene. The mother cow became acutely aware of a predator in her midst and made a sudden movement toward him. Kuri, always quick-witted and with a strong instinct for self-preservation, rightly judged the threat of angry hooves and withdrew to a safe distance. Frank took advantage of this distraction to retrieve the stillborn calf. When the mother returned she soon directed her attention to Tiny Tim and began to lick the placental slime off his slight, waiflike body. Later that day, Frank carried the dead calf, which he estimated to weigh sixty-five pounds, into the woods behind the pasture and buried it under a pile of stones.

Monday rolled around. Frank returned to his carpentry job in New York City and our energies went back to growing vegetables. It was a hot, dry summer and I remember spending lots of time setting up irrigation lines and tasting the dryness of dust in my mouth. Sometime later that week I noticed that Aldo, the male Maremma, was not eating his dog food. Each morning, his partially full bowl of crunchies was attracting starlings and other scavenging birds who invariably peppered any remaining food with their droppings. It was a matter of some concern. Was there something wrong with Aldo? Might he be sick? To all appearances, he seemed his normal, contented self-if anything, more contented than usual. We considered changing brands of dog food or switching to wet, canned food for a while, but Tiki, his sister, was quite approving of what we gave them.

By Friday the answer to Aldo's appetite shift appeared on the back lawn. It was the stillborn calf—or rather, some small remaining portion of it. Apparently, Aldo had gone into the woods, uncovered the carcass, and eaten about three-quarters of it in place. When it got small enough for him to drag, he must have decided it would be more secure if he brought it back to the house. What remained was not much more than the skull and spinal column with a little surrounding flesh, but Aldo guarded it fiercely. Whenever Tiki approached too closely, no doubt hoping that her brother might show some sibling generosity, Aldo, flat on the ground with a jaw full of veal bone, emitted deep growling sounds that persuaded her to turn away.

Frank Sgroi and I have a good symbiotic relationship. I am happy to have his cows visit each year to dine on our summer pasture and Frank is happy to bring them. His own heavily wooded land is not adequate to satisfy their appetites and he would otherwise be forced to provide them with large amounts of purchased hay and other cattle feeds. The grass and assorted browse in our pasture is healthy, low-cost food. It's a good deal for Frank, but with one caveat. He has to maintain some several thousand feet of old barbed wire fence so that the cows will stay in the pasture allotted to them and not range into the surrounding woods or, worse yet, our vegetable fields.

Unfortunately, the fences seem to always be in need of repair. They are easily damaged by falling trees or cows pushing their heads through to reach greener grass on the other side. Every year numerous breakouts occur, and since Frank is usually in New York City, it falls to me and my crew to round up the culprits and herd them back into the pasture. Depending on the number of animals loose and their degree of friskiness (young heifers are the worst), this can be a challenging task. At the very least, it requires coordinated teamwork and a careful, deliberate approach. If the cows get spooked and run in the wrong direction, they can cover a lot of ground very quickly and wreak much damage along the way. This has happened more than once, to my considerable dismay.

An escaped cow eats with five mouths and the most voracious are the four that point straight downward. Any vegetable that feels

I've always been fond of animals . . . Compared to my own species, they have much to admire.

the imprint of a cow's hooves is finished, at least as a saleable item. One of my more witty interns once informed me that the cows had broken through the fence that morning and creamed our onions. At the time, I didn't find much humor in the remark. In the midst of a busy picking day we managed to round up the fugitives and funnel them back into the pasture. I parked a tractor and manure spreader by the ruptured fence to foil further breakouts and I think I can recall leaving an angry message on Frank's answering machine, which I may later have regretted. After all, an old dairy farm deserves to have some cows, and cows have wills of their own, and the farming life is one of challenge and uncertainty. And it must be said, in their defense, that the cows contribute much, including the aura of their presence, their manure in the pasture, and their meat on the dinner table—all probably worth a good deal more than a few creamed onions.

FARM POLITIC

More than money, energy is what makes the world go round. Of this you can be sure, by being hungry. It all begins with the sun's vast reservoir of electromagnetic energy, which radiates out into space. Next comes the ability of plants to capture some of that energy through photosynthesis and convert it into carbohydrates, the basic food for many of the earth's higher life forms.

Food is the fuel that gets us up in the morning and keeps us going throughout the day. It's what enables us to function and prosper and grow. The quality of the food we eat determines the quality of our health and well-being. Though we live in a land with a large

The nation's
best soils are
becoming
increasingly
lifeless,
degraded, and
vulnerable to
erosion.

agricultural capacity and a surplus of inexpensive food, not all of that food is good for our health. Most of us appear to be amply fed, and clearly many are more than amply fed. But size alone is a poor measure of health. First-time visitors to the United States are often taken aback by the girth and waddling gait of many of our citizens.

In the United States, a large share of our agricultural efforts are focused on just three food crops: wheat, corn, and soybeans. These crops, like all others, capture the sun's energy, but, unlike fruits and vegetables, which are usually eaten almost in their entirety, wheat, corn, and soybeans are harvested primarily for their seeds or grains. The seeds of these three plants have a particular knack for concentrating the sun's energy and storing it in a stable form. This attribute makes it possible for agribusiness companies to accumulate large quantities of wheat, corn, and soybeans, and then process them into other foods or trade them nationally and globally. In this way wheat, corn, and soybeans become commodities or forms of wealth.

To grow these three commodity crops, you need adequate rainfall and a suitable climate. You also need fertile soil. Unfortunately, we're facing a limited and diminishing supply of this third critical resource. Intensive industrial farming practices have robbed many soils of their natural fertility. Since the end of World War II we've overcome this problem by using increasing quantities of petrochemical fertilizer and pesticides. These synthetic inputs have kept crop yields up, but have taken a toll on the environment.

The nation's best soils are becoming increasingly lifeless, degraded, and vulnerable to erosion. We've lost about half of our topsoil to erosion in the last fifty years and continue to lose soil at a rate seventeen times faster than nature can replace it. Rivers, lakes, and coastal waters that receive runoff from farm fields are not faring any better. Nearly all the nation's surface waters in agricultural areas are contaminated with nitrogen, phosphorus, and an assortment of pesticides. The Mississippi River, which drains the nation's agricultural heartland, carries a chemical soup that has created a dead zone the size of New Jersey in the once biologically rich Gulf of Mexico.

Fossil fuels, primarily oil and coal, represent the harvest of the sun from eons ago and as such are highly concentrated forms of energy. It is not surprising, then, that they should be the chief ingredient in chemical fertilizer—which itself is a concentrated form of energy, as Timothy McVeigh, with his homemade fertilizer bomb, clearly demonstrated.

Very large amounts of oil or coal are used in the production of agricultural chemicals. It takes approximately four hundred pounds of coal to produce one pound of nitrogenous fertilizer, and many millions of tons of fertilizer are used every year. Subtract most of the fossil fuel from the modern agricultural equation, and farmers would have to go back to organic methods. It might not be such a bad thing. The Cubans have, of necessity, had considerable success in this area, but I doubt we'll be emulating them anytime soon.

We in the United States have a deep love affair with fossil fuel, especially when it comes in the form of oil. We consume oil at a per capita rate far greater than any other nation. We range the earth in search of this finite and ever-diminishing resource. And we've made it clear to the rest of the earth's inhabitants that we are willing to embark upon unprovoked foreign wars and pro- longed military occupations to ensure that our oil needs are met. As we sit down to eat our evening meal, we might reflect that it is not only the wheels of industry and the wheels of our motor vehi- cles that turn with oil. Our agribusiness, our national food econ- omy, is very, very oil dependent.

Each calorie of food we consume typically has behind it at least a calorie's worth of oil, and often many more. In the case of highly processed foods, the amount of fossil fuel energy needed can be quite staggering—on the order of ten or more calories' worth of fuel to produce one calorie of food. You might assume that all that energy would make for better food. Ironically, there is often an inverse relationship between the amount of fossil fuel energy used in the production of a given food item and that food's ability to maintain a healthy human body. Invariably, processing strips the natural goodness out of food. Often it becomes downright unhealthy. Two such examples of processed foods ubiquitous in our daily lives are hydrogenated oils and high-fructose corn syrup.

Hydrogenated oils, or trans fats, are manufactured by a process that forces hydrogen into oil, usually soybean oil, at high temperatures. The resulting product is thicker than oil and confers a longer shelf life to the many foods that contain it. This commonly consumed additive may indulge your taste buds, but it's not good for your health.

High-fructose corn syrup, a processed derivative of corn, is used to sweeten just about everything, from soft drinks to cookies to ketchup. About three-quarters of all processed foods contain it. High-fructose corn syrup both caters to and increases our craving for sweet things, but is practically devoid of nutritional value. It does, however, contribute to diabetes, obesity, and heart disease—all ailments that are afflicting us at epidemic levels.

Check the labels on canned and packaged food and beverages in your supermarket shopping cart sometime. You might be surprised at how much hydrogenated oil and high-fructose corn syrup you're consuming. Then hop on a scale and check your weight. Here the relationship between the one and the other is not likely to be inverse.

Our agricultural oil habit extends deeply into the meat industry as well. Most of the beef, pork, and chicken we consume comes from animals that are raised largely on commodity grains, which, as already noted, are grown with massive petrochemical input. As many as thirty-five calories of fossil fuel are needed to produce one calorie of industrial beef. Indeed, the bulk of the corn and soy grown in this country goes to fatten livestock in corporate feedlots, despite the fact that cattle, hogs, and chickens are most happy and healthy when left to graze and forage for themselves on the open range.

There are still other ways in which our food industry consumes prodigious amounts of energy. Long-distance transportation and refrigeration are fairly obvious examples. Less well known

Our agricultural oil habit extends deeply into the meat industry as well.

is modern agriculture's heavy dependence on petroleum-based plastics. Every year, millions of acres of farmland are covered with plastics to warm the soil and suppress weeds. Most irrigation systems also rely heavily on plastic products. At the end of each season, most of these plastics end up in our landfills.

The 2002 Farm Bill that made its way through Congress and was signed by President Bush contains extremely generous subsidies for the largest corporate farms. The lion's share of the $13.4 billion paid out in 2006 went, as always, to the major commodity crops—corn, wheat, cotton, soybeans and rice. Over 75 percent of subsidies were granted to the top 10 percent of agricultural producers, widening the gulf between small and large farms. To get some idea of the scale of farm subsidies, consider Riceland Foods, of Stuttgart, Arizona, growers of rice, soybeans, wheat, and corn. One of the biggest beneficiaries of federal largesse in 2004, Riceland received over $7.7 million. That must have been a bit of a disappointment for them, considering that over the twelve-year period between 1995 and 2006 Riceland pocketed a total of $554,343,039 in tax-payer funded subsidies. To learn more about the big winners in the farm subsidy game, visit the Web site of the Environmental Working Group. You'll notice that less than one percent of the Farm Bill's huge subsidy allocation goes to assist fruit and vegetable growers.

To small, diversified farmers like myself, this skewed distribution of funds seems a lot like corporate welfare. And, in our free market society, it creates an uneven playing field. Subsidies make corn, wheat, and soybeans unnaturally cheap and encourage overproduction, which leads to surpluses, which bring prices down even further. Large buyers and processors of grain reap the benefits. Meanwhile, unsubsidized, and therefore more expensive, crops are naturally less attractive to consumers. No wonder small farmers, who are capable of producing more natural, fresh, and wholesome food using less energy, are becoming an endangered species. And no wonder our nation is plagued with obesity—given all the processed, surplus grain we are being fed.

To small, diversified farmers like myself, this skewed distribution of funds seems a lot like corporate welfare.

So we have a corporate food system that is based heavily on three commodity grains; that depends on fossil fuel energy almost every step of the way; that impoverishes our farmland and farm communities; and that takes a heavy toll on the larger environment. Much of the food generated by this industrial system goes to feed livestock in cramped, cruel, and unhealthy conditions where large doses of antibiotics and other medications are essential. Much of the rest is processed beyond recognition into "junk" foods that are virtually devoid of nutritional value and often harmful to our health. Because these faux foods are designed to taste good and are vigorously promoted through advertising, they are widely appealing, especially to children. They also engender addictive cravings. All this is hardly a recipe for national health.

How did we get to this state of affairs, you might ask? In my view, we got here by placing too much faith in a corporate, profit-oriented system—a system that puts stock prices, executive compensation, and market dominance well above the public's interest. When it comes to food, health, and environment, this is a flawed model, made ever more alarming by the trend toward massive consolidation within the food, seed, and agricultural chemical industries. The fact that this system is supported and underwritten by federal and state authorities makes it all the more egregious. Through favoritism, subsidies, and unfair regulations, our government has helped to create a food system that is controlled by a handful of corporate giants with little or no accountability.

It has become exceedingly difficult to draw a clear line between the big corporations and the governmental agencies (the USDA, FDA, and EPA) that are supposed to regulate them. Via revolving doors, individuals at the highest levels move from industry to government and back again with impunity. In a very real way, these people are determining what the nation's diet will consist of. And there can be little doubt that their ambitions extend well beyond our national borders.

As a small vegetable and herb grower, I'm an advocate of a

different kind of agriculture. I'd like to see a food system that is less dependent on wheat, corn, and soybeans, and less focused on international trade and the global economy.

I'd like to see more small farms, more local production, less monoculture, less chemical dependency, and more unprocessed, decent, wholesome food. I'd like to see more focus on fresh fruits and vegetables. At the very least it would help us to keep our weight down and most likely keep health-care costs down, too.

I'd like to see us wean ourselves off fossil fuel as much as possible. I'd like to see fewer feedlots and more grass-fed livestock. I'd like to see a type of agriculture that really belongs to people, rather than to Archer Daniels Midland, ConAgra, Cargill, Monsanto, Altria and the like. I'd like to see our government get out of the way.

THE MONSTER IN THE CLOSET

WHILE I'M LAYING out my wish list, there's one more thing I'd like to see, and that is a concerted global effort to address the monster in the closet—population growth. Without taking much notice, we have nearly doubled our numbers to 6.4 billion humans in the last forty years. Currently, we are adding to our ranks at the rate of 200,000 people every single day. That's a net global increase of 1 million every five days. The earth's capacity to support humans is not infinite. Each of us needs a livable environment, plus shelter, food, and water. The pressure on these resources is already intense. It's doubtful that our planet could sustain its current human population if every person alive today consumed as much as the average American.

And there is the small matter of the myriad other species who have made the earth their home and have lived alongside us since our beginnings. They, too, need an environment to inhabit and food to eat. Might we not find within ourselves the generosity of heart to leave a little space for them?

At the beginning of the twentieth century there were about 100,000 tigers still roaming the wilds of Asia. Today, optimistic estimates put the number at around 3,500. Two subspecies—the Caspian tiger and the Javan tiger—became extinct sometime in the 1970s while I was playing poker and driving a cab in New York City. I wasn't aware of their passing at the time, but now, thirty-five years later, the knowledge has settled in. Like shadows in the night, they stalk me wherever I go.

IN THE END it will come down to the will and wisdom of the people. Do we wish to shape our own destiny, or are we willing to leave it in the hands of others, who often benefit mightily at our expense? Do we care about future generations, or is it just us, here and now, who matter? Do the words *stewardship* and *sustainability* have any resonance in our lives, or are they just two outmoded, sentimental, and commercially regressive concepts?

In the small-farm world that I inhabit, stewardship is a privilege and a responsibility to be nurtured and encouraged and not conflated with personal gain. Sustainability is a way of thinking about and practicing farming that assumes future generations will rely on the same land to feed and nourish themselves. Sadly, it has proven difficult to practice this style of agriculture at the modern corporate level.

But at the grassroots level, there is change in the air. I meet more and more people who share the same vision of a local, sustainable, humane, and vibrant agriculture. Many flourishing examples can be found, including grow-your-own farmers' markets, food cooperatives, urban rooftop gardens, and CSAs, especially in and around our major cities. (CSA stands for Community Supported Agriculture.) Members of a CSA agree to share in the risks, rewards, and vagaries of a diversified farming operation. Before the season begins, a CSA farmer sells weekly shares of his or her anticipated harvest. As the season unfolds, a bountiful harvest means members will receive larger weekly shares; a meager harvest, be it due to bad weather, pests, or plant disease, but hopefully not the negligence of the farmer, means the reverse. In the CSA model, the farmer gets paid for his or her efforts whether the outcome is fruitful or not. Some CSAs encourage member participation in the farming process, especially in the harvest and the distribution of the weekly food packages.) More than just a few people are beginning to understand that systemic change is needed. Legions of others are finding themselves increasingly dissatisfied with the processed industrial food that is offered to them. Across the nation, a host of groups and individuals are working for positive change. More are needed. It won't be easy. But when was a fight that's really worth winning easy?

KURI—
circa 1985 to 2003

I*t has been* more than two years since Kuri died and I am ready to tell his story. The incident that led to his death happened on a sunny Friday afternoon in May. I remember it well. The week had been productive. My new crew was settling in nicely and in good spirits, no doubt looking forward to a couple of days off. The late-afternoon sun cast a soft light on the newly planted fields, highlighting the work we had done; the trees along the driveway stood half-clad in their new, tender green foliage. In the distance the surface of the pond was glistening. All seemed well, and I was pleased with our progress for the week though still hurrying to get a few more chores done.

Kuri had been taking it easy; I hadn't seen much of him. Gone were the days when he would follow me out to the field and spend hours by my side or settle down in the shade of a nearby tree and wait for me. He was a very old dog, and I was under no illusions about the state of his health. We all knew his days were numbered. For months I had been anticipating his end—preparing myself, I suppose.

Longtime regular visitors to the farm—the UPS delivery man, the heating-oil truck driver, the woman from the post office who brings packages too big for the mailbox to our front door—would often ask after him. He had been around so long, and most of them regarded him fondly.

"How's the old dog?" they would ask, if he didn't come out to meet them. Or, "I see he's not getting around too well." "Lost his bark, has he?"

Sometimes they'd come over to give him a pat on the head and a dog biscuit, and see his big, crooked tail wag slowly from side to side. They remembered the days when he would come running toward them, a little menacingly, with his sharp bark intended to alert me to their presence and to serve as a warning that foul play would not be tolerated. But that spring, even a dog biscuit was too much for Kuri, too brittle and hard for his nearly toothless mouth to chew. He'd wander off and drop it in some high grass at the edge of the lawn, where one of the other dogs, usually Aldo, would claim it.

Against considerable odds, Kuri had survived the winter of 2003. Through the coldest nights he held on to his life with admirable tenacity. Though in a state of advanced decrepitude and presumably a fair amount of pain, it was clear that he would not go willingly into death. When spring came it was like a reprieve. I was proud of him. He had made it through in his own indomitable way. Perhaps he would stay with me for another season.

But I found myself constantly wondering how his end would come. Would he suffer increasing pain? Would he deteriorate to the point that we might have to put him down? If so, would I have the heart to do it myself? Or would I take him to some foreign place and hand him over to a vet in a white coat with a syringe in his hand?

Which would he prefer? I'm glad I did not have to make that hard choice, though, as it turned out, I was to be the agent of his death.

The year before, we had built a large, greenhouselike structure that we refer to as "the high tunnel." It has two four-foot-wide swinging doors at either end and roll-up sides, but no fan and no heater. During cool days, if you keep the doors closed and the sides down, the temperature inside can be ten or fifteen degrees warmer than outside. On sunny days, if it gets too hot, you can easily raise the sides and open the doors to let cooler air flow through. The purpose of this tunnel is to extend our season a few weeks in both spring and fall. We plant directly into the ground, as we would in the field, but are able to offer the plants some protection from extremes of weather.

My plan that spring was to plant an early crop of basil in the tunnel. But first I had to go in with my Allis Chalmers tractor and a small rototiller and prepare a seed bed. It was the last thing on my list for that Friday. Once I got the tiller on the tractor it took only ten or fifteen minutes to complete the job. I was exiting very slowly through the wide front doors. It was a difficult maneuver. There were only two or three inches of clearance on either side. The bucket in front of the tractor partially obscured my view, and instead of looking forward, I was focused on the right back wheel, making sure it did not hit the door frame. Just as I cleared the door I felt the tractor ride over a large soft bump on the ground that I knew shouldn't have been there.

Already with a sinking feeling in my stomach, I stopped the tractor and got off. There was Kuri lying behind the left rear wheel. His body was shaking and he appeared to be gasping for breath. I knelt down beside him and gently touched his face and spoke his name. He did not respond, but continued to shake and started to make short coughing noises. Fighting back tears, I called to one of my workers and asked him to bring me a canvas tarpaulin. We gently spread the tarp under Kuri and carried him into the shade of the tractor shed. I sprinkled cool water on his face, but he barely responded.

Before long, my wife, Flavia, arrived home. Together we decided to take Kuri to an animal hospital in Goshen, hoping there might be some chance he could survive the crushing weight I had brought down upon him. At about seven o'clock on Friday night, there were still several people in the hospital waiting room with their cats and dogs and assorted other small creatures. The staff encouraged us to leave Kuri with them overnight and go home. They assured us they would give him the best care they could. I reluctantly agreed, knowing in my heart that it would probably be the last time I'd see him alive. And it was. The next morning they called to tell us that he had gone during the night. They said it was probably for the best, given his advanced age and the severity of his injury.

And so it was Kuri's last act of friendship—to come and wait for me at the end of the day, even if it was a great effort for him to do so. Probably he had fallen asleep while I worked in the tunnel and, being deaf, did not hear the tractor as it slowly moved toward him. Had I not been in such a hurry, might I have been more careful? Might I have thought to look out for him? Who can say? But there's one thing I do know: Sometimes in life our smallest actions bring about radical change that cannot be reversed. There's a verse from the Rubaiyat of Omar Khayyám that my mother was fond of:

The moving finger writes; and, having writ
Moves on: nor all your piety nor wit,
Shall lure it back to cancel half a line,
Nor all your tears wash out a word of it.

With the passage of time, I have become reconciled to what happened on that pleasant afternoon in May. Perhaps it was for the best. Perhaps it was fitting that, at the end of his long and eventful life, Kuri should die by my own hand. In a way it cemented the bond between us.

Of all the animals I have known, and there have been many, my attachment to Kuri was the strongest. He was my first dog and will always occupy a large place in my heart. He came uninvited in the early days of the farm. I turned him away more than once, then took him in with much uncertainty, doubting he would contribute practical value to my already shaky endeavors and fearing another dimension of responsibility. He quickly won me over, however, and introduced me to his intense canine world. Looking back to those early days, it seems that Kuri came to us with clear intent, that he purposely selected our farm to live on and perhaps us to live with. Abandoned and alone in the world, he found his way to a new home.

Kuri was a more complicated dog than most and he had something of a split personality. He alternated, on an almost daily basis, between loyal friend and guardian of our well-being on the one hand, and free agent, canine hunter and adventurer, on the other. When returning from his forays into the woods and swamps of the surrounding countryside, he always had a wild-eyed look to him, as though he had experienced some heightened physical state. Sometimes I was greatly displeased with his roving ways and the troubles they occasionally caused me, but a small part of me always secretly relished them. It was his engagement in the living world and his periodic uninhibited wildness that most endeared him to me.

In his younger days, Kuri was able to emit a most wonderful, long, melancholy howl, though I rarely heard him do it. According to my wife, he invariably howled for extended periods of time whenever I left the farm in my blue pickup truck. When he heard the truck returning, even before it was visible to him, he would come bounding forward with his head and tail aloft, eager to be the first to welcome me home. As I got out of the truck, he would nudge my legs with his nose several times by way of greeting and to register any foreign smells I might have brought with me.

In midlife, to our chagrin, Kuri lost his ability to howl. This happened almost certainly as a result of his castration—a procedure I unhappily agreed to, though one that, ironically, almost certainly extended his life by seven or eight years.

In those days our neighbor to the south was a part-time backyard breeder of Doberman pinschers. It was a rough and unruly

In his younger days, Kuri was able to emit a most wonderful, long, melancholy howl.

operation with a half-dozen muscular and highly aggressive males chained up in the yard at all times and a few females who often ran free but, to my knowledge, stayed within the confines of the neighbor's property.

Back then, before housing developments came to pockmark the rolling landscape, Kuri roamed at will, far and wide. Sometimes he was sighted by friends a mile or two from our house. I've no doubt he regarded the neighbor's property, at least the wooded and open sections of it, as part of his range and quite naturally would have been drawn there whenever he sensed receptive females of his own species.

One day the neighbor with the Dobermans, whom I had met only once and briefly, came down our driveway. He didn't get out of his car, but I recognized who he was and went over to greet him. Immediately I saw that he was sweating profusely and in an agitated state. Without so much as a "hello," he blurted out: "Your dog's been fucking my bitches."

At a loss for words, I broke into an awkward but spontaneous burst of laughter. This did little to improve matters. By now Kuri was at my side. The angry neighbor pointed a stubby finger at him.

"You chain him up or I'll slap you with a lawsuit." He almost spat the words at me. Then, as an afterthought, he added, "Or get him castrated."

Still somewhat taken aback by what I was hearing, and thinking to inject some levity into what was, to my mind, a rather absurd and essentially comical exchange, I looked at Kuri, who was showing keen interest in what was taking place, and asked him directly, "How do you feel about that, Kuri? Should I tie you up for the rest of your life or cut your balls off?"

Evidently, I had further misjudged the situation. The neighbor became almost apoplectic at my irreverent tone. For a moment it looked like he might get out of his car and accost me. But perhaps he reconsidered upon reflecting that he was noticeably older and smaller than I (with a bald head and potbelly to prove it). At any rate, I was confident that a physical threat could be easily dealt

And so it was Kuri's last act of friendship— to come and wait for me at the end of the day, even if it was a great effort for him to do so.

with. And, had it come to that, I've no doubt Kuri would have been delighted to employ his teeth on my behalf.

After a few more nasty words about lawsuits and my being in violation of a town ordinance requiring that dogs be leashed at all times and not permitted to run freely on other people's property (an ordinance that was probably rarely enforced) the neighbor drove off, taking his ill humor and poorly developed social skills with him.

Over the next few days my wife and I discussed Kuri's fate at length. I was against castration, while she was leaning toward it. The threatened lawsuit was not a matter of great concern to us. We believed that the neighbor himself was in violation of a town ordinance for having a commercial dog-breeding business within three hundred feet of our property line. So we had our own threat to level at him, if need be. But we had to acknowledge that Kuri's insemination of other people's female dogs (which, judging by the neighbor's irate performance, had already occurred, along with resulting progeny) was indeed a problem, though, with respect to the neighbor's Dobermans, I was inclined to believe that any action there on Kuri's part could only result in a distinct improvement to the gene pool.

We also had to admit to a fear that Kuri's free-ranging lifestyle, while it clearly suited him, might lead to even more serious problems. We knew he had a propensity for chasing cats and chickens. What else might he determine to be fair game?

My wife spoke to a couple of vets who told her that castration might tame him down and reduce his wandering habits (I don't think it did), and even increase his life span. They argued moreover that it was the responsible thing to do, given the large number of unwanted puppies in the world and resulting high rates of canine euthanasia. What choice did I have? Moreover, I felt that Kuri would opt for continued freedom over confinement, even at the high cost of castration. After all, what good would his testicles do him if he had to spend the rest of his days restrained on our front lawn? A decision was taken: For Kuri's own good, and ours, his balls would have to go.

It's a Long Road
to a Tomato
252

An appointment was made, and Kuri, who had never visited a vet while in our care (we had taken him a few times to an annual rabies clinic, for prophylactic shots), cooperated fully, always delighted to be taken for a ride in a vehicle. That evening we picked him up at the vet's office and brought him back to the farm. He was a little groggy at first but seemed relieved to be back on familiar ground. I detected no great concern on his part for the loss he had suffered, though for a couple of hours he did concentrate a lot of licking energy on the spot where his testicles had been.

Two days later the vet called to tell us that Kuri had heartworm—a parasitic condition, common in free-ranging dogs, in which worms penetrate and feed on their host's heart. He estimated that, without treatment, Kuri would not live beyond another year. We took him to another vet and had him retested—the results were the same. Thus, we were faced with another hard decision. The treatment, we were told, would be expensive—over $1,000. It would take place over a few months, consisting of several doses of arsenic and overnight stays at a veterinary hospital. It could also be dangerous, perhaps even fatal. But we decided it was worth the risk.

The heartworm treatment was indeed stressful, both for Kuri and for us. During that time he suffered a great loss of energy and some hair and spent many hours sleeping in the barn. Even months after the treatment was concluded there were side effects, most notably a general listlessness and persistent skin rashes. For a while we wondered about the wisdom of our actions. Would it have been better to let nature take its course? With the passage of time, the answer proved to be a resounding no. In less than a year Kuri was back in fine fettle, ranging across the fields, chasing raccoons and woodchucks, and generally exhibiting his great enthusiasm for life. We were much relieved and knew we had done right by him. He stayed with us for seven more years, through the prime of his life, and his being there added immeasurably to the character and spirit of the farm.

Looking back at his time with us, I can recall many days on which Kuri brought both drama and humor into our lives. One of

His being there added immeasurably to the character and spirit of the farm.

the more alarming and ultimately absurd occurred when Aldo and Tiki were puppies, still learning the ways of the world. By nature, these two are guard dogs, satisfied to protect a defined territory and those they determine to be under their care. But being young and impressionable, they were greatly influenced by Kuri, whose interests were far more wide-ranging. His passion was for the hunt, and the company of other dogs, even a couple of furry white youngsters like Aldo and Tiki, only increased his daring. He was undoubtedly the leader of the pack and, against my wishes, he often took the puppies with him.

One morning when I left the house I noticed that all three dogs were gone. While this was unusual, at this time of day, I didn't pay much heed, assuming they would be back before long. Besides, it was a Friday, and we had plenty of harvesting to do for a late-summer Saturday market. Through the morning I was vaguely aware of their absence but stayed focused on the work at hand. By one o'clock, when I came in for lunch, the dogs were still nowhere to be seen. I began to feel some alarm. Kuri, I felt sure, could take care of himself and find his way back, but the puppies could easily get lost or struck on the road. Forgoing my customary midday nap, I drove and walked to the edges of the farm and yelled the dogs' names at the top of my lungs. I then left the farm and spent nearly half an hour driving nearby roads searching for them, but with no success.

That afternoon we continued picking for market, but I grew increasingly anxious and angry. I knew Kuri was to blame and felt he should have known better. I imagined how I would chastise him and tie him up for at least a day or two. He had to learn that this kind of behavior was unacceptable. But still they didn't come back. By 4:30 PM I felt a pressing need to do something. I went into the house and called several neighbors, and then a couple of animal shelters to find out if any stray dogs had been sighted. Finally, with some trepidation, I called the state police, who have a station just a couple of miles from our farm. I spoke to a trooper Hannigan, and gave him my telephone number and address and a description of the dogs. He reminded me of the town's leash law,

Kuri in His Grave

to which I feigned ignorance, but said he would put out an alert and help me try to find my dogs. What more could I do?

My wife was staying overnight in New York City and I had to get up at three o'clock in the morning, eat breakfast, pack a lunch, and leave for market. It was a bad situation and I anticipated a short and worried night of sleep.

By 7:30 PM, we had finished loading the truck and I was walking back to the house when, to my great relief, I spotted Aldo in the half light. I called to him but he moved away, perhaps detecting some anger and distress in my tone. Soon after, Tiki emerged from the shadows, and finally Kuri. For a few moments they kept their distance, as though needing time to adjust to domestic order after their lawless activity. The puppies had mud on their faces and underparts and looked bedraggled, but Kuri was his usual sprightly self. Finally I got a grip on him and dragged him to his doghouse, where I chained him for the night. I'm sure he knew that he had displeased me greatly and accepted his punishment as a reasonable price to pay for a day on the lam. I spoke harshly to Aldo and Tiki but doubt they understood what wrongs they had committed. I ate a quick dinner and went to bed totally exhausted.

The next thing I knew the phone by the bed was ringing. It roused me from a deep sleep. As I struggled to pick up the receiver, I saw by the light of a full moon that the time on my alarm clock was five minutes to two. What asshole was calling me at this hour of the night? Or might it be an emergency? It turned out to be Trooper Hannigan. He asked me if my dogs had returned. I couldn't believe that he was calling me at nearly 2:00 AM with such a question. I answered him curtly: "Yes, they did," and then mentioned that it was nearly two o'clock in the morning. He responded, "It's eleven-ten PM" I said, "It's five minutes to two, for God's sake, and I was sound asleep." We each repeated ourselves once more, and then his tone changed. He spoke slowly and deliberately and with an air of officialdom.

"Mr. Stewart," he said. "It is ten minutes past eleven on Friday night and I am about to end my shift. I'm sorry if I woke you up. But I haven't heard from you and I need to know if you found your dogs so I can fill out my report for the day." Something told me that I ought not disagree with him any further. I apologized for not calling (I had completely forgotten), thanked him for his concern, and told him again that the dogs had returned home. Each of us then bade the other a short goodnight.

I then turned on the light and noticed that it was indeed 11:10 PM or rather, by then, nearly 11:15 PM. Being abruptly wakened from deep sleep and relying only on the light of the moon, I had confused the hour and minute hands of my clock, which are almost the same length, to arrive at 1:55 AM. I felt another surge of anger toward Kuri, and then the sheer absurdity of what had just transpired dawned on me and I began to laugh out loud. What sort of a nut did Trooper Hannigan think I was?

◆

YEARS LATER, ON that Saturday morning in May, we retrieved Kuri's body from the vet's office, avoiding the faces of the several dog and cat owners who were happily conversing in the parking lot. The staff had taped him tightly inside a black plastic bag, which seemed to me unnatural and terribly confining. Back at the farm I cut him loose and let him lay in the back of my truck for a few hours so that his stiffening body might absorb the sounds and smells of his home.

That afternoon I dug his grave near a grove of yew trees behind the house. We wrapped him gently in one of my old shirts and laid him in the ground. His eyes were closed, his back legs were pulled up against his body, and his front paws were bent close to his face. He looked like he often did when sleeping on a cold night, trying to hold on to all the warmth he could. I ran my fingers along his nose and up to his forehead one last time, bade him farewell, and pulled the loose dirt over him with my hands. Tears rolled down my face and onto the earth that would now enfold him. I knew then that I had lost as good a friend as I would ever have, one who threw in his lot with me and attached himself to this rough patch of land with a commitment and fervor similar to my own.

BREAKDOWN:
perils of the
truck-farming life

It *was* 4:30 AM on a cold morning in December. I don't know how long the phone had been ringing, but it was long enough to rouse me from a deep, dreamless sleep. Suddenly I was talking to Ben, one of my employees. His voice was faint. I kept asking him to repeat himself, perhaps because I'm somewhat deaf and didn't have my hearing aids, or perhaps because I was not quite ready to register what he was telling me. Finally, I had to accept the relevant words: "Keith, the truck has broken down."

A sinking feeling set in. All I wanted to do was go back to bed and reclaim the sleep I had been so rudely deprived of. But that option was rapidly fading. I began to assemble a few relevant details. My Wednesday market crew—Steph, Ben, and Monica—were in the middle of an open stretch of County Route 1, just about fifteen minutes from the farm. There was no shoulder. The truck had seemed low on power from the beginning of the trip, especially when going up hills. Then it suddenly died. They had the headlights on so other vehicles would not rear-end them, but had no heat and it was extremely cold. I got Ben's cell phone number and promised to get back to them as soon as I could.

Downstairs in the kitchen, I searched the Yellow Pages for local tow truck companies. My first call woke an elderly woman who, once fully conscious, told me in a weary voice, from very far away, that there would be no drivers available for at least three hours. I felt bad for having woken her; I could definitely relate. After a couple more calls I found a place in the Town of Goshen that said they would send out a tow truck right away.

I called Ben and told him that I'd be over to get them in twenty minutes. I just needed time to go to the bathroom and get dressed. I was now fully awake and feeling a little better. At least we had a plan.

Five minutes later I stepped out into the frigid air. My first inhalation felt like an icepack in the lungs. Later, I learned that the mercury that night had descended to minus nine degrees Fahrenheit, which turned out to be the lowest temperature of the entire winter—a cold night, to be sure. Even a brass monkey would tell you that.

I jumped into my Ford pickup and worked the key into the ignition. The engine turned over sluggishly but would not start. I tried a few more times, but to no avail. For a few minutes I sat in a state of chilly despair, longing only for the warmth of my bed and the sweet, undemanding oblivion of sleep. Then I thought of the big battery charger I had purchased at a yard sale a year or two before but had not yet found a use for. Perhaps it was worth trying. In the dark shed, I located the heavy charger and wheeled it over to the truck. I found an extension cord. With frozen fingers, I hooked up the terminals

I stepped out into the frigid air. My first inhalation felt like an icepack in the lungs.

I've had my
share of truck
breakdowns
and a bit of
bad luck with
tractors as
well.

and set the device on "Engine Crank," the highest of four settings. I then ran back inside, made myself a cup of tea, and called Ben. Through a fading connection, he told me that the tow truck had just arrived and would transport the three of them to a gas station in the nearby town of Pine Island. Then the phone went dead.

I waited five minutes before returning to the pickup truck. This time, to my great relief, the engine started. I disconnected the charger, drove to Pine Island, and found my crew in an all-night convenience store. They were clutching bags of garlic and shallots they had salvaged from the market truck—to prevent them from freezing. Bundled up against the cold, they looked bizarrely out of place under fluorescent lighting, between aisles of brightly colored candy and soda. We threw the alliums into the back of the pickup, crammed into the cab, and drove back to the farm. It was still dark outside and not many houses had their lights on. Though a little shaken by the experience, my workers seemed quite relieved not to have to spend such a cold day in the streets of Manhattan, peddling freezing vegetables. I can't say that I blamed them.

Later that afternoon I learned that the subzero temperatures had caused the diesel fuel in my Mitsubishi truck to gel and burst the fuel filter. It could have been avoided, I was told, if I'd added an antigel fluid to the fuel tank the night before. Too bad I hadn't known that.

We truck farmers are very dependent on our trucks. They take us to market where we exchange our goods for hard cash to pay the bills and stay in business. A broken-down truck filled with fresh food can bring about a bipolar mood swing in no time. All that effort for nothing. A precious day of marketing lost, and often a good amount of perishable food lost too. There's also the hassle of getting your large, disabled vehicle to a repair shop and transporting everyone back to the farm, not to mention the considerable costs involved.

Over the years, I've had my share of truck breakdowns and a bit of bad luck with tractors as well. I wish I could say I've become more philosophical and learned to accept these incidents as part and parcel of the farming life. But that's not the case; I still live in dread of breakdowns and fervently hope I will somehow elude my allotted quota of mechanical misfortune. The fact that I've always been a somewhat mechanically challenged individual, except at the most rudimentary level, surely adds to the vulnerability I feel.

If you had the choice, you'd always take a truck breakdown at the end of day rather than at the beginning. At least then you've converted your load into negotiable currency, which definitely helps. But, on the downside, after a day at market you're probably on the verge of exhaustion and in no condition to tackle a vehicular crisis. The normal nonstop, high-demand, seventeen-hour day can morph into something a lot more stressful.

My first big truck breakdown happened about fifteen years ago. I was heading north on 10th Avenue in Manhattan, with a couple of helpers, after a particularly hot Saturday market at Union Square. We were stopped at a red light when a passerby shouted that there was smoke coming from under the engine. I pulled over and got out to take a look and, sure enough, he was right. But what were we to do? Abandon the vehicle right there on 10th Avenue and let it burn? Pray that the smoke might go away? Amazingly, while I was trying to decide on the right course of action, the smoke did subside. We got back in, a little uneasily to be sure, and proceeded on our route to the Lincoln Tunnel. But not for long.

The problem, I was to find out later, was a leaky master cylinder dripping brake fluid onto the exhaust manifold, whence the smoke arose. The vehicle, apparently, was in no great danger of catching fire. More troubling was what we experienced next: brake failure— a consequence of the lost brake fluid. I gave up on trying to reach the Tunnel and drove, very slowly, with one hand on the parking brake, to the nearest garage. There, I reluctantly left the truck in the care of a non-English-speaking attendant.

One of my workers then managed to persuade a friend in Manhattan to come to the rescue and drive us back to the farm in his own vehicle. That was much appreciated. On the ride home we got to laugh and make light of our recent misfortune. "It's an ill wind that blows no good" is a wise expression from my homeland.

Once again, it turned out to be true—our rescuer, a hearty fellow by the name of Ivan, ended up coming to work for me the following year and was a fine addition to the farm.

On Monday morning I caught a bus to the city and, a few hours later, retook possession of my truck, with a new master cylinder installed.

A couple of years later, that same truck—an aged, one-ton GMC with a sixteen-foot box—broke down again, after market, on a desolate stretch of Route 17 in New Jersey. It cruised to a halt at a red light and refused to go any further. None of the three of us on board had a cell phone and there were no open businesses in sight, so we simply turned on the hazard lights and stood outside waiting, assuming a patrol car would appear before long. None did, but within several minutes a tow truck stopped alongside us, as though the driver had known in advance that we would come to a halt at that very spot. He leaned out his window and offered, for an exorbitant fee, to tow us to an all-night garage a few miles away.

Reluctantly, I agreed, but when we arrived at the destination, I wondered about the wisdom of doing so. The place was an exceedingly rough and dingy establishment in an otherwise abandoned area. Moreover, it seemed to be minimally equipped. Any description of the proprietor of this "garage" and his lurking associates might have put the word "unsavory" to good use. Thoughts of the New Jersey mafia crossed my mind. That my pockets were bulging with some $3,000 in cash (much of it in small bills) probably magnified the sense of unease I was feeling. But there we were and, short of running into the street, there wasn't much we could do to alter our fate. On the positive side, my two assistants were healthy young men, reasonably well-endowed with muscle and testosterone.

The three of us strode about confidently and occasionally congregated in a corner of the building to converse in gruff tones. A couple of hours passed during which time the proprietor and one of his assistants intermittently examined various parts of the engine with a drop light. I was beginning to feel that we were in no particular danger, but I doubted that we would get back to the farm that night.

Then, somewhere before midnight, the man walked over to me and said that he had discovered the problem. It was the rotor—a component I had never heard of. He opened a metal cabinet, pulled out a circular plastic disk a few inches in diameter, placed it in the right spot under the hood and then confidently climbed into the cab and started the truck. When he got out he was smiling and didn't look so unsavory any more. Suddenly he was our savior. I thanked him, though still affecting a gruff tone of voice, and willingly paid him the few hundred dollars he asked for. Then we set off into the night. My sense of relief was so great that the remaining hour and a half of the journey didn't seem that long at all.

The third breakdown in that particular truck occurred at the farm on a Saturday morning. I simply couldn't get the thing started, no matter what I tried. I even called a mechanic friend at 5:00 in the morning. He came over but was no more successful than I. After this occasion, on which a fair amount of fresh food was lost, I decided to get rid of that truck. It was costing me too much money.

My current market vehicle, the Mitsubishi, has served me well enough for ten years. There was, of course, the incident with the frozen fuel filter, but that particular debacle can be attributed more to the ignorance of the owner and the coldness of the night, rather than any defect in the machine itself. Certainly, other repairs have been needed, but none so urgent that a day at market was lost.

Last year, though, before I uninsured the vehicle for the winter, there were a couple of unnerving incidents. The Mitsubishi developed an intermittent problem with either the starter or the solenoid. My mechanic had not been able to determine which and was reluctant to replace both of them. So, for a while, I was stuck with the problem: On occasion the truck refused to start, using normal methods.

A couple of times I was able to get the engine running by crossing terminals on the starter with a screw driver. This causes sparks to fly and is a little dicey. If you forget to put your transmission into neutral, when attempting this maneuver, you could get run over by your own vehicle. One morning, before I knew this trick with the screw driver, I drove into the Union Square market area and turned off the

engine prior to parking in my designated space. I did this so that we could easily remove the long poles that frame our canopy. When I tried to start the engine a few minutes later all I got was a little click. The problem was solved only after I had assembled a group of farmers to push the truck ten or fifteen feet into its proper space.

The rest of the day went well. We sold most of our load but I couldn't stop worrying about how I would get the truck started for the ride home. I mentioned our predicament to a few neighboring farmers, hoping one of them might come up with some useful advice. The most encouraging response came from my friend, Andy Van Glad, who sells maple syrup next to us on Saturdays. He licked his upper lip, expanded his already substantial chest and said: "Don't worry, Keith, we'll get you home." I felt better already.

At the end of the day, Andy borrowed a stout though rather worn and tattered piece of rope from another farmer. His plan was to use his truck to pull mine until we reached sufficient speed to effect a jump-start. We tied the rope to the front axle of my truck

It's a Long Road
to a Tomato
264

and the rear of his and waited till the market cleared out a bit and the farmer in front of Andy left for the day. Still, we didn't have much room to execute the move—maybe 120 clear feet in front of Andy's truck. There were still a few pedestrians in the market so my helpers were assigned the task of making sure that none walked blindly in front of us.

Inside the Mitsubishi, I turned the ignition switch to ON, shifted into second gear, kept my foot on the clutch, then signaled to Andy. We started to roll, slowly at first, then faster. Knowing we had only one good chance to make this work, I decided to wait until the very last second. We were moving at a fairly good clip when an elderly woman ambled dangerously close to my left mirror, then another, younger one shouted, at the top of her voice, "Your rope is breaking!" When I heard that, I lifted my foot off the clutch. There was a sudden jerk as the Mitsubishi went from neutral to second gear and as Andy's truck made its last violent pull. The rope snapped and whipped to one side. A couple of pedestrians jumped back. I swiftly put one foot back on the clutch, the other on the brake, and came to a halt. Then I heard the sound of the Mitsubishi's motor purring. Ah, what a sweet sound it was.

Over the years, truck troubles have definitely bruised my psyche and may have contributed to the graying of my beard. But the market truck has not been my only source of mechanical anguish. As a farmer, I'm also heavily dependent on tractors, and I could recount several incidents in which I have suffered on their account as well. But I'd like to end on a more positive note.

Last summer, a mishap involving my largest tractor, while at the time a source of embarrassment for me, turned out to be a real blessing in disguise. After a couple of hours of planting winter squash, I had brought my John Deere 4040 to rest in a pole barn near our house. Attached to the back of the 4040 was a water wheel transplanter with two eighty-gallon tanks that had just been replenished with water. The combined weight of the rig was probably in the neighborhood of twelve tons. In hindsight, it would have been wise to take this into account before going inside to use

I heard the sound of the Mitsubishi's motor purring. Ah, what a sweet sound it was.

*Mishap with the
Big Tractor*

the bathroom, especially since the transplanter was extending outside the pole barn a good ten feet, in a downhill direction.

Barely a minute had passed when my wife, who was on the phone, and apparently looking out the window, emitted a loud shriek, indicating that she wanted my attention immediately. I came running to see what was wrong, but was too late to witness the cause of her alarm. In a state of shock, she simply pointed in the direction of the pole barn.

When I looked out the window, the first thing I noticed was that the tractor and transplanter were not where I had left them. Instead, they were seriously engaged with another building—a cinderblock tool shed—seventy-five feet down the slope from the pole barn. What I really mean to say is that the tractor and transplanter had, under the inescapable force of gravity, left the pole barn, rolled down the slope, and caved in one end of the other building. There was a large hole in the shed, where the transplanter had entered. Cinderblocks were lying on the ground. The front of the structure and its roof were pitching forward perilously.

It dawned on me that I had forgotten to put the tractor in park—embarrassing, yes, but otherwise I felt no great remorse.

Neither the tractor nor the transplanter had suffered significant damage and no human injuries were sustained—always a good thing. The tool shed, however, was finished. No doubt about that. But the truth is, I disliked that shed intensely. It was a dingy, crumbling place with a low ceiling and a lower doorway that I had repeatedly knocked my head on, causing considerable pain and giving rise to permanent scars on my bald spot. It was also full of junk that really needed to go to the dump. And it was home to various rodents, to boot.

Later that summer, my master-builder friend, Tom Berg, along with his son Michael, demolished what remained of the cinder-block shed and put a far superior structure in its place. This new shed has a much higher ceiling, red metal siding, a concrete floor, a sliding door, and many desirable interior features. It is a truly great workshop. Every time I enter it I feel quite pleased with myself and glad that I had the foresight, on that summer day, not to put the big tractor in park.

ABOUT SEEDS

That *diminutive speck* of genetic knowledge and prodigious energy we call a "seed" will, when given time and suitable conditions, transform itself into a six-foot-tall tomato plant weighed down with sumptuous fruits; a tight green head of broccoli rich in vitamins, minerals, and anti-oxidants; a purple plum radish radiant with color after its coating of soil is washed away; or myriad other earthly delights. It's worth pausing, once in a while, to ponder this miraculous feat—life's pivotal ability to replicate itself, with a little help from the sun and soil.

In the short, cold days of winter, vegetable growers, like myself, revisit the world of seeds. We turn to catalogs, scribbled notes, past experience, and tips from other growers. There are so many choices, so many seductive possibilities to explore. On our farm, we grow over a hundred different varieties of vegetables and herbs. Each has its own set of traits, its desirable qualities, and sometimes regrettable shortcomings.

Should flavor be our foremost concern? What about productivity, rate of growth, length of harvest period, resistance to pests and diseases, frost tolerance, tolerance to heat and drought, shape, size, color, past performance, nutritional value, transportability, storability? Or that other rather large concern for a commercial grower: marketplace appeal.

There are also questions about a seed's origin—where it comes from, how it was grown, the extent to which it has been subjected to human manipulation. Is the seed in question an open-pollinated variety, a hybrid, an heirloom? Was it harvested from organically grown plants? Or is it a new, genetically modified and patented seed, under the exclusive dominion of a large bio-tech company?

Certified organic growers must use organically grown seed when it is available in the varieties we choose to plant. When not available, conventionally produced seed may be used, but never seed that is the result of genetic manipulation. The organic seed industry has come a long way in the last ten years but still represents only a small percentage of total seeds sold in the United States. Nevertheless, today, more than 80 percent of the seed I use comes from organically grown plants—that's a lot more than just a few years back. Organic seed almost always costs more.

Open-pollinated plants are largely a product of evolution and selection of preferred traits by humans. They breed true to type and have viable seed. This means that a gardener or grower can save and replant seeds from open-pollinated plants and expect offspring just like the parent. There's one caveat though: If you're planning to save seed from an open-pollinated vegetable, you may need to isolate it from other varieties of the same species. Otherwise, you could get unexpected crosses. Take winter squash, for example—Acorn,

In the short, cold days of winter, vegetable growers, like myself, revisit the world of seeds.

Delicata, and Sweet Dumpling are three distinct varieties of *Cucurbita pepo* (*Cucurbita* being the genus and *pepo* the species). If grown within the same field without any physical barriers between them, there's a good chance that these three tasty squashes will cross-pollinate with each other—the results will be woody, gourd-like fruits with little culinary value.

Hybrids, often referred to as F1 hybrids (which stands for first filial generation), are a little different from open-pollinated seeds. They are obtained by deliberately cross-fertilizing, in a controlled environment, two or more parents from different varieties of the same or two closely related species. If the hybridization is successful, the result will be a new and often superior plant that shares some of the positive traits of each parent. The problem with most hybrid vegetables is that they're good for just one generation. In the second generation their seeds are nearly always sterile or at least unpredictable in performance. Gardeners with even a modicum of knowledge don't waste their time saving hybrid seeds. Each year they buy them from a seed company and usually pay more than they would for open-pollinated varieties.

Though often developed by universities and government research stations, hybrid seeds usually end up under the control of large seed companies. They may have some superior characteristics, especially when it comes to uniformity and shelf life, but if a seed company decides to drop a particular hybrid, that variety is lost to the public, even though it may have performed well. For several years I planted Emperor, a hybrid variety of broccoli, which grew impressively big heads, tasted good, and was well received by my customers. Then one year I noticed it was not in the seed catalogs I order from. When I called to find out why, I was told it had been discontinued by the large European seed company that had developed it. Several seasons have gone by and I've yet to find a broccoli I like as much.

Many farmers and gardeners worry that hybrids will supplant open-pollinated seeds which, if not propagated regularly, tend to disappear. If you're a fan of small, self-sufficient farms and genetic diversity, you probably don't want to see the world totally overrun with hybrids. For this reason, organic growers tend to favor open-pollinated varieties. But sometimes hybrids are just too good to pass up—the white Japanese turnip called Hakurei is one example. If I stopped growing these crisp, sweet turnips, especially delicious when eaten raw, simply because they are hybrid, my customers would probably rise up in revolt. The same might be said of the Sungold tomato—an amazingly sweet, orange cherry tomato that also happens to be a result of hybridization. About one quarter of the seeds planted on our farm last season were hybrids. The rest were open-pollinated. This ratio has been fairly constant for several years.

Seed saving, like most worthwhile endeavors, takes time, knowledge, and patience, and it can be a little risky for the novice since there is some danger of carrying a seed-borne plant disease from one year to the next. I've dabbled in seed saving now and then but mostly have declined, feeling that I lack the time and expertise to do a good job. But when it comes to garlic, I make an exception. I'm very attached to the Rocambole variety which has been growing on our farm for twenty years and is well-adapted to the local soils and climate. Each year we set aside about three-quarters of a ton of garlic bulbs, which will be divided into cloves, for planting stock. In years of bountiful harvest we may also save potatoes and shallots for future planting.

Fortunately there are some intrepid gardeners and a few farmers who still go to the considerable effort of saving their favorite varieties of open-pollinated seeds. There are also organizations dedicated to preserving a genetically diverse, open-pollinated seed bank. Foremost among these in the United States is the Seed Savers Exchange in Decorah, Iowa, which was founded in 1975 by Diane Ott Whealy and Kent Whealy. Shortly before his death, Diane's grandfather passed on to Diane the seeds of a German Pink tomato and a Morning Glory plant. His own parents had carried earlier generations of these seeds with them when they left Bavaria in the 1870s to immigrate to America. Her grandfather's act provided the inspiration for starting the Seed Savers Exchange. Today

Seed saving, like most worthwhile endeavors, takes time, knowledge, and patience.

Germinating Seed

the organization has over 9,000 members and their yearbook lists more than 13,000 varieties of vegetables, fruits, and grains, which are available to members. Seed Savers' primary mission is to propagate and distribute open-pollinated, heirloom seeds that are at risk of being lost forever. The USDA also maintains seed banks of open-pollinated vegetable varieties for research purposes.

Heirlooms are open-pollinated plants that have been grown by humans for at least fifty years, and often much longer. They are often, but not necessarily, produced organically. Many heirlooms have a history of being passed down by gardeners or farmers from one generation to another, outside of the confines of the commercial seed industry. More than most plants, they reside in the public domain.

Many heirlooms were brought to this country by early European settlers (old varieties of cabbage, carrot, and parsnip come to mind) or by more recent immigrants. Two wonderful heirloom kales on our farm are Lacinato (also known as Black Tuscan or Dinosaur) Kale, from Italy, and Red Russian Kale, which was first introduced to Canada by Russian traders.

Some heirloom vegetables, especially certain varieties of corn, beans, and squash, were planted and cultivated by Native Americans long before Europeans arrived on this continent. One notable variety is a snap pole bean called "Cherokee Trail of Tears." These small black beans were carried by ancestors of the Cherokee Indians on the infamous death march in the winter of 1838-39, when the Cherokee were forcibly relocated from North Carolina to

Oklahoma. Some 4,000 of them perished along the way. For the past couple of years we've grown a small patch of Cherokee Trail of Tears beans in our backyard. They're pretty good, either as dried beans or while still tender and in the pod. Last year my workers ate all of them.

For a small truck farmer like me, heirlooms can be a good bet. They are definitely in vogue these days. At my Union Square market in Manhattan, business in heirlooms has been brisk for several years and shows no sign of abating. Heirloom tomatoes are especially popular. Brandywine, Cherokee Purple, Black Krim, and Striped German are a few of my favorites. The fruits of these heirlooms are usually large, richly colored, and variable in shape. Flavors vary but most are exceptional. A good heirloom, ripened in hot sun, can be everything you ever dreamed a tomato could be, both in taste and appearance.

On the down side, many heirloom tomatoes are thin-skinned and have an unfortunate tendency to bruise or split, especially when hit with heavy rain in the days before harvest. Moreover, they can ripen unevenly and tend to be poor travelers. For these reasons you're unlikely to come across heirloom tomatoes in a supermarket. And if you do find them in an upscale food store, you can be fairly sure they were picked well before ripeness. But for small farmers like me, these sometimes irksome traits are blessings in disguise. They keep the heirloom tomato business out of the clutches of the industrial food system and leave it to us. There's little doubt that heirloom tomatoes are best suited to backyard production and local truck farming.

In the world of plants, genetically modified (GM or GMO) seeds are a very different bag of beans. Proponents and practitioners of genetic modification will tell you that they are just following in a long tradition of human plant breeding and selection. What they don't tell you is that their gene-splicing techniques are unlike any form of propagation that has taken place before.

Genetic modification of plants does not rely on the traditional reproductive methods which have been used to create hybrids and

ABOUT SEEDS

273

open-pollinated seeds in the past. These traditional methods involve sexual breeding and crossing of different plant varieties within the same or closely related species. For example, a wild onion with a very small bulb but excellent disease resistance might be hybridized with a large-bulbed, domestic onion which is not, strictly speaking, the same species. After some trial and error, the result could be a domestic onion of large size and improved disease resistance.

Most transgenic seeds are created by using one of two decidedly nonsexual approaches. Currently the most common method uses infectious bacteria to insert foreign genes into a plant's cells. The second, older, but perhaps more quintessentially American approach uses a gene gun (a sort of air gun) to shoot tiny particles of gold, which are coated with foreign genetic material, into the cells of the subject plant.

Another thing that sets GMO seeds apart is the dissimilarity of the life forms that are forcibly conjoined. An early example was Calgene's Flavr Savr Tomato in which a gene from a flounder was spliced into a tomato. The intent was to create a tomato that could tolerate colder temperatures and presumably withstand several days of refrigeration in a tractor trailer without loss of flavor. The buying public, however, was not impressed with this coupling of a tomato and a flounder, the progeny of which didn't taste particularly good. Despite much advance publicity, the experiment ended up floundering badly.

The Calgene Tomato may have decomposed on the ocean floor, but other transgenic crops are already entrenched in our food supply. These days, if you buy supermarket products containing corn or soy (which is most of them), you're almost certainly ingesting genetically modified food. In 2009, Monsanto's Round-Up-Ready Soybeans accounted for 85 percent of the soybeans grown in the United States. As mentioned in the chapter "Brave New Vegetables," these soybeans are engineered to withstand Monsanto's broad-spectrum herbicide, Round-Up, which kills most leafy plants it comes in contact with. This "miracle" of bio-engineering has done wonders for Roundup sales.

Ninety percent of the corn grown in the United States in 2009 was also subjected to genetic manipulation, having had the foreign bacterium *Bacillus thuriengensis* (Bt) inserted into the cell structure of the plant in order to control an insect called the European corn borer. What effect this newfound insect toxin might have on humans and the environment is not well understood, but a study published by the Austrian government in 2008 suggests some ominous possibilities. After feeding GMO corn to multiple generations of mice, researchers noticed that their subjects had weakened kidneys, decreased fertility, and inflammation and cholesterol problems. To me, using Bt corn seems a bit like going after a mosquito with a jackhammer. If you're fast, you'll kill the insect, but you might end up putting a hole through your wall, or your arm, in the process.

Many scientists have expressed concern that these new transgenic organisms, which the world has literally never known before, are being released into the environment with almost no long-term testing or government oversight, yet the USDA, presumably under pressure from the biotech industry, has decided that GMO foods do not need to be labeled as such. This means we don't know when we're eating them, unless we choose 100 percent certified organic products which, by law, can not contain GMO ingredients.

There is also the question of ownership. As soon as they are developed, GMO crops are patented as proprietary, intellectual property. Unlike open-pollinated and heirloom seeds, they are definitely not in the public domain. They are invariably owned and controlled by a handful of large corporations (Monsanto, Dupont, and Syngenta are the biggest) who charge farmers ever-higher prices for the seeds they have created. Today, these three companies and a few others are racing to dominate the global food market. They have spent billions developing and promoting their products and they expect to get paid back, many times over, regardless of what it means for the rest of us.

Lately, the GMO giants have stepped up their "green fields" ad campaigns and found some surprising voices to defend them, like

I'll stick with my heirlooms and open-pollinated seeds and maybe a few good tasting, old-fashioned hybrids.

Stewart Brand of "Whole Earth Catalog" fame and Michael Specter, a staff writer for *The New Yorker*. The looming crises we and our planet face, most notably climate change and population growth, are used to silence critics. Anyone who questions GMOs is accused of backward-thinking anti-scientism, or of being indifferent to the world's starving masses—in hot, dry weather, no less. That's a convenient and self-serving position to take if your goal is to control the earth's food supply. But it's not corroborated by fact or scientific study. A report published in 2009 by the Union of Concerned Scientists clearly states that, after thirteen years of commercial use, GMO seeds have not led to significant increases in crop yield, despite industry claims to the contrary. What they have led to is significant increases in profits and consolidation among agro-tech companies. So much is about seizing the market.

Do we really want to entrust our food system to a few big transnationals, especially in this era of lax government oversight and corporate dominance? Are we willing to let them determine what we will eat and how it will be grown, without having any obligation to tell us the details? Despite reassuring TV commercials, showing amber fields of grain, smiling faces, and the promise of abundant food for a hungry world, do we really believe that these giants have our best interests at heart? To assume they do seems, to me, dangerously naive. Until persuaded otherwise, I'll stick with my heirlooms and open-pollinated seeds and maybe a few good tasting, old-fashioned hybrids—while I still can.

A BEAVER
before
BREAKFAST

've just come inside to catch my breath after chasing a fine-looking beaver across a couple of fields and into the pond. What a way to start the day. Certainly, I've encountered a few beavers in my time—usually with much delight—but never before have I seen a specimen such as this, up close, on dry land, and on this farm.

I had walked to the greenhouse before breakfast to check on the several thousand seedlings planted over the previous few weeks. The night had been a cold one for April. There was a dusting of snow on the ground and the outside temperature was not much above freezing, but the sun was shining in a clear sky and it was pleasantly warm under the protective plastic of the greenhouse. I set about misting several flats of germinating parsley seeds that had spent the night on heat mats and were in need of a little moisture.

Tiki, our female Maremma and loyal, protective companion, had come over to the greenhouse with me, but stayed outside, as usual, in her guard dog capacity. A couple of minutes later she began to bark. I didn't pay much heed, since most of her barking is preemptive, designed to let would-be intruders know that she is on duty and alert. Shortly, though, I noticed a shrillness and urgency to her vocal emanations that suggested she might be onto something more substantial than a figment of her imagination. I set down the watering wand and left the greenhouse to take a look.

There, in the grass, not twenty feet away, looking straight at me, or more likely at Tiki, was a large beast with a reddish brown coat, small eyes, and short, rounded ears. At first I mistook it for a woodchuck, though one of such massive girth that it might have dined on my entire previous season's crop of butternut squash, or else was a holdover from the Pleistocene megafauna of some ten or fifteen thousand years ago when oversized mammals roamed the land. But soon I noticed that the beast, which was backed up against a hedgerow and clearly in a defensive posture, had a tail the size and shape of a squash paddle. It suddenly occurred to me that I was staring into the eyes of a beaver.

Perhaps emboldened by my presence, Tiki ratcheted up her barking a few decibels and moved closer to her quarry, albeit in a circumlocutory fashion, but looking very much like she was bent on attack. Had it been a woodchuck, the creature would probably have been lodged between her powerful jaws by then, and I might have witnessed only the final, violent, spine-shattering twists of her head. But this was something entirely different and much bigger. Prudence dictated a more cautious and strategic approach.

And, indeed, had it been a woodchuck, the natural enemy of every vegetable grower, I might have stood by and allowed Tiki to satisfy her predatory desires. But, for me, this was a rare and wondrous sight—a full-grown beaver, dark-coated and glistening in the

It suddenly occurred to me that I was staring into the eyes of a beaver.

early sun, backed up against a tree, its life in the balance. I was not about to stand by and watch this impressive creature get torn apart by my dog, especially not on the premise that she was protecting me. Further, I'll admit to some concern for Tiki's well-being—I did not doubt she would prevail in the end, given her advantage in size, her swifter legs, and her predatory inclinations, but I feared she might incur some injury in the process. Any animal that can fell a good-sized tree with its teeth has got to have some serious biting power. Out of concern and for their mutual well-being, I stepped into the fray.

First I tried to get hold of Tiki's collar, but she eluded me. Then I barked a couple of sharp "No!"s at her. She hesitated for a moment and looked back in my direction with unbelieving eyes. "Surely you don't disapprove," she seemed to be saying. The beaver, meanwhile, stood its ground, doubtless aware that to run and expose its flank to this white, wolflike canine would be a grave error. Tiki, not so easily dissuaded, quickly returned her attention to the furry monster. She resumed barking and moved to within five or six feet of the beaver, keeping a low profile, as though preparing to strike.

I saw there was no time to waste and ran toward her, yelling, "Off, off!"—a command she is familiar with and which usually brings about the desired result. This time she pulled back long enough for me to grab her collar. But her will to persist was strong; she struggled vigorously against me, repeatedly lunging forward with her teeth bared. I was scarcely able to restrain her eighty-plus pounds of muscle, sinew, and intent. Meanwhile, the beaver, presumably mystified by what was going on, but perhaps detecting some improvement in the situation, moved a few steps sideways and positioned itself in front of another, smaller tree. It continued, however, to face Tiki frontally, watching her every move.

Tiki then changed her strategy. Instead of straining forward in the direction of the beaver, she began to pull backwards, against me, twisting her head from side to side in an effort to free herself from the collar. In this she almost succeeded, for the dogs' collars are never so tight that they are unable to extricate themselves in an emergency. I understood what she was trying to do and stepped back, lifting her by the head as I did. In this manner I managed to keep my hold, all the while shouting, "Off, off!" in her ear. Gradually she accepted my rightful dominance and allowed me to pull her away in the direction of the tractor shed. The beaver, noticing a greater distance between itself and this bizarre, wrestling twosome, began to move off in an awkward, waddling gait.

Soon I had Tiki in the tractor shed. I clipped a chain to her collar and she sat down next to her brother, Aldo, who was not quite done with his night's rest. Aldo, though larger than Tiki, is a more passive and gentle soul. He was either oblivious of, or unconcerned by, the drama that had just unfolded.

With Tiki secure, I left the two dogs and ran back into the field, eager to reengage the beaver. By then it was mid-way through a patch of last year's broccoli stubble, moving in the direction of the pond. Its progress was painfully slow and ponderous for an animal whose life was so recently in peril. With its stubby front legs and extra-wide body, it was clearly not suited to rapid movement on land. Before long I was at its side, just a few feet away, marveling at its girth and extraordinary tail. Unaware that I harbored no ill intent but was merely trying to encourage its movement in the direction of the pond and enjoy a little interspecies exchange in the process, the beaver tried to evade me by desperately lunging forward, using its broad, flat tail to lift itself off the ground. This was impressive to witness but hardly effective. It managed to gain an elevation of only six or nine inches before it came down head first, at which point it briefly lost its footing and almost rolled over on itself. In one of these awkward tumbles, I noticed that the beast's back legs were much sturdier and its feet seriously webbed.

Concerned that my looming presence might cause the beaver to injure itself or perhaps suffer a coronary mishap, I pulled back and allowed the animal to proceed under less harried circumstances. It continued to head in the direction of the pond, though from such a low vantage point that it would have been impossible to see the water. How then did it know which way to go? Was it familiar with our pond and the surrounding terrain? I seriously doubted that. Or was it able to sniff out the whereabouts of water? Of course, it was heading downhill, which would be wise for any creature hoping to find a pond or stream. But there were other downhill directions that it could have taken where no water lay.

Eventually the beaver did reach the banks of the pond, whereupon it immediately dived in, with some small amount of grace, and disappeared. A minute later it surfaced roughly in the middle of the pond and, coincidentally, within a few feet of a pair of Canada geese. The geese seemed quite unnerved by this large and unusual intruder in their waters and quickly paddled away, their necks straining forward in the direction of travel.

The beaver, now clearly in its element, proceeded to swim with surprising speed around the perimeter of the pond, which is about a half acre in size. Only its nose, eyes, and perhaps ears appeared above the surface, but I could easily see the rippling

The geese seemed quite unnerved by this large and unusual intruder in their waters and quickly paddled away.

swell created by that wonderful tail which appeared to be setting the animal's course.

I surmised that the beaver was either conducting some form of reconnaissance in this new body of water, or else demonstrating its aquatic prowess and, in the process, making it clear to me that I would no longer have any chance of catching it. This aquatic display lasted for a few minutes then abruptly ended when the beaver's tail rose up in the air and came down with a tremendous thwack, causing a great splash of water. The two geese were clearly upset by this loud, sharp sound; they swam directly to the edge of the pond and climbed out as fast as they could. The beaver, meanwhile, disappeared into the watery depths, perhaps to conduct subsurface investigations.

Though not well-clad for such a chilly morning, I stayed on the banks of the pond for several minutes, waiting to see if the beaver would reemerge. It didn't, or, if it did, it kept its nose well hidden along the reedy edges. Confident that the creature was no longer in danger, I returned to the tractor shed and removed the chain from Tiki's collar. Accompanied by Aldo, she returned to the scene of her earlier adventure and, using her nose, proceeded to trace the beaver's course across the fields and right up to the bank of the pond, at the exact point where the subject had dived in. Aldo, now equally animated, followed her. When the two of them arrived at the water's edge, they stopped and looked out over it for several moments, perhaps wistfully reflecting on the big one that got away.

*C*ASTOR CANADENSIS—THE MORE formal name bestowed upon the beaver—is the largest of North American rodents (a beaver can weigh as much as sixty pounds) and usually lives in a colony or family unit with several others of its kind. Beavers are generally monogamous and mate for life but they will "remarry" if a spouse dies or otherwise disappears. Juvenile beavers stay with their parents for approximately two years before they are asked to leave. Cast out into the larger world, it is up to the young animals to find mates and build lodges of their own. It is reasonable to assume that the individual I encountered that morning was such a one, wandering in search of a new home. Beyond our northwest border there's a good patch of wetland which is probably where this beaver came from.

Beavers are largely nocturnal and are seldom sighted during daylight hours. They are amphibious; their webbed back feet and rudderlike tails give them good mobility in the watery realm they mostly inhabit. They dine on tree bark and cambium—the moist, nutrient-rich layer just behind the bark.

As we all know, beavers like to keep themselves busy. They are skilled dam builders and construct and maintain elaborate, partially submerged lodges for their residential use. Trees, because they provide both food and building materials, are pivotal to the beavers' survival—especially such species as willow, birch, maple, aspen, and poplar. An experienced beaver with a good set of front teeth can fell a five-inch-diameter willow in three minutes flat. For the most part, beavers take down smaller trees which can be thinned from the forest with little, if any, negative impact. The dams and ponds they create provide valuable habitat for many other wildlife species, but are not always so warmly received by humans.

The beaver's thick, well-insulated coat, highly sought after by humans, was almost its undoing. In the early nineteenth century, beaver pelts constituted the single most important trade commodity in much of North America. Most ended up in east coast cities or European capitals where they were fashioned into items of attire. Great fortunes and financial empires, some of which persist to this day, were made from the backs of beavers. Eventually, the trapping of beavers reached such a rapacious level that the animal was virtually eliminated from much of its range. More recently, efforts to reintroduce and protect beavers are beginning to pay off. Though they are still not common in the Hudson Valley, there is yet a chance that you too, on some clear spring morning, with a dusting of snow on the ground, will have the good fortune to encounter a beaver before breakfast.

An experienced beaver with a good set of front teeth can fell a five-inch-diameter willow in three minutes flat.

THE EVEN LONGER ROAD to a TOMATO
or
THE RAIN IT RAINETH EVERY DAY

(Midsummer 2009)

couple of weeks ago I paid a neighbor with an excavator to dig a hole, six feet deep, in some fallow land behind the pond. Then, with grim resolve, my crew and I pulled several hundred dying tomato plants from the field—some already had green fruits on them that were turning brown. We threw them into the hole and covered them with a layer of soil. It was a painful thing to do.

Before that, I had been advised to destroy most of the shallots we had planted in April and some of the onions. Now, I'm looking at the potatoes and wondering what their fate will be.

Total rainfall on our farm in the month of June was 12.1 inches. The average is around four. One Hudson Valley weather station logged 13.66 inches, the most rain in June in more than a hundred years. It was definitely a wet month, and a chilly one. Everyone knows that plants need water in order to grow, and rain falling out of the sky is essentially free water that irrigates everything at once. So why should a farmer complain? Ever heard the expression "too much of a good thing"?

Under normal growing conditions, plant roots penetrate deep into the soil (as deep as two or three feet) looking for moisture. If the upper layer of the soil is constantly wet, plants have no incentive to send roots down prospecting for water. Shallow-rooted plants have less access to nutrients and minerals and, if conditions turn dry, their roots are not well situated to capture moisture at deeper levels. Therefore, too much water can lead to weaker, malnourished plants. Excessive and heavy rain can also result in loss of topsoil through erosion. No farmer wants to see that happen. And too much rain can cause nutrients in the topsoil, especially nitrogen and potassium, to leach down to deeper levels where they are no longer accessible to plants.

But there's another, more sinister, reason why too many rainy days can lead to trouble, and that is the proliferation of disease. Fungal and bacterial diseases are ever-present in the plant world, but they are mostly held in check by the intricate ecology of natural systems. The damage they do is usually not severe. But give them rain, day after day, week after week, and it's an entirely different story.

The first sign that excessive rainfall and the large number of rainy days in June was causing trouble showed up in our shallots. Near the end of the month, I got a phone call from an upstate farmer with whom I had shared an order of shallot sets, which had been shipped down from Canada at considerable expense. (Shallots can be grown from seeds or "sets." Sets are little bulblets—as

Plants that
had been
growing
happily a day
or two earlier
looked like
they had just
been hit with a
blowtorch.

they grow, they divide into clusters of new shallots). He asked how my shallots were doing and told me that the leaves on his were turning black. I put the phone down and went straight out to inspect my crop and, sure enough, plants that had been growing happily a day or two earlier looked like they had just been hit with a blowtorch. The tops of their leaves were shriveling, turning black, and dying. It was quite a shock.

I called my local Cooperative Extension agent, Maire Ullrich, who has an Agriculture degree from Cornell and twenty years of field experience, and has frequently advised me in the past. Maire deals with trouble. She maintains, with some humor, that no one ever calls her when things are going well. She came out to the farm the next day, took one look, and said "Downy Mildew," pointing out the grayish spores in the middle of the dark patches on the shallot leaves.

Maire then added that I had earned the distinction of having only the second case of Downy Mildew in alliums that she had seen in Orange County in her twenty professional years. She assumed it had come to our farm on the imported Canadian planting stock, especially since my farmer friend in the north was having the same problem (he later confirmed that his shallots were also afflicted with Downy Mildew). The damp, cool weather, Maire said, would have enabled the disease to take hold and spread on both of our farms.

I stood there, absorbing the bad news and thinking of the many days we had already invested in planting and weeding those shallots, not to mention their initial cost of over $500. Maire, meanwhile, moved away and cast her trained eye elsewhere. Before I had regained full composure, she pointed out that Downy Mildew was already surfacing in some nearby onions. Her advice: get rid of both the infected shallots and the onions. Plow them in, without delay. Then start spraying a fungicide on all the other alliums (onions, shallots, scallions, garlic) or risk losing the whole lot.

There are not many fungicides approved for organic growers, and those that are don't get high marks for effectiveness. I had no

idea what to use. Maire consulted with some Cornell disease experts and, a day or two later, came up with a product called OxiDate, which is essentially a very concentrated and expensive form of hydrogen peroxide. I obtained a five-gallon bucket of the stuff for $295. I then went out and bought a tractor-mounted boom sprayer and asked one of my workers to assemble the thing and figure out how to use it.

I began spraying all the alliums that showed no signs, or minimal signs, of disease. Meanwhile, my crew went to work cutting down and covering the several thousand row-feet of infected shallots that we had already put so much work into. Due to all the rain, the ground was too wet for plowing so we opted to cover the diseased plants rather than incorporate them into the soil. I waited a few days on the onions and, when conditions were a little dryer, reluctantly turned them in with a rototiller.

Dealing with this sudden outbreak of disease took its toll on my (and probably my workers') reserves of psychic capital. And it used up a lot of valuable time—time that might have been spent on other chores such as planting, weeding, mulching, trellising, and generally tending the numerous other crops we grow, many of which had not received the attention they should have, due to all the rainy days in June.

On top of this, spraying pesticides does not fall into the category of fun jobs. Not by a long shot. Even organically approved sprays can be dangerous—they must be used with precision and care. Too much can damage the crop; not enough may have little positive effect. The label on the jug of OxiDate warned of blindness should any of the concentrate find its way into your eyes, and death should you make the mistake of drinking it. I donned plastic-lined overalls, Nitrile gloves, goggles, and a respirator, none of which feel good on a hot day. And I kept my mouth shut.

There also was a larger concern. In addition to attacking disease, fungicides may negatively affect other organisms in the air or soil. What industrious microbe, beneficial or otherwise, wants to be doused with hydrogen peroxide while going about its business on a sunny day? As I sprayed, I worried, too, about the barn swallows flying overhead, dining, opportunistically, on the tiny fungicide-coated insects that were flushed up as the tractor and spray boom traversed the fields. Worse still, I imagined that many of these insects were destined to go into the waiting mouths of baby swallows still in their nests in the barn.

Nothing about spraying is appealing, except the hoped-for, positive end result. But what's a farmer to do? Lose the crop he or she has already spent months working on? Lose the anticipated income? Disappoint loyal customers? I can hear them now, by the hundreds, asking why we don't have shallots, onions, and—I can barely summon the courage to say it—garlic, at our stand this year. It's a rare farmer, organic or otherwise, who will not, sooner or later, face tough choices like these. As it turned out, we had more tough choices coming our way. The next problem surfaced a week later, in both our tomatoes and potatoes, and it was worse.

Both of these crops had been progressing nicely. Our first planting of tomatoes had already flowered and set fruit, and an acre of potato plants were flowering aboveground and developing tubers below. I might lose most of the shallots and onions, I told myself, but a good harvest of potatoes and, especially, tomatoes would help make up for the loss. That was like counting chickens before they hatched.

While trellising tomato plants on the first of July, I noticed the occasional olive-brown splotch on a few leaves. Probably nothing to worry about, I thought. Tomatoes contract all manner of diseases but still manage to bear marketable fruit. In my twenty-two years of growing them, I'd suffered a few setbacks to be sure (Early Blight, Blossom End Rot, Bacterial Speck, and Spot) but had always had a fair number of good-looking and good-tasting tomatoes to sell. I was eager, however, to hear Maire Ullrich's assessment of my spraying program for the alliums, so decided to give her another call.

She came out the next day and, again, it didn't take long for her to make a pronouncement: "You've got it, Keith. Late blight! Sorry." I shuddered inwardly as I remembered hearing, a day or two before, late blight referred to as the "bubonic plague of plant diseases."

I knew, from reading and talking to other growers, that, within a week, healthy plants can become shriveled remnants of their former selves. At first, brown lesions appear on leaves and stems. Soon after, dark, leather-like spots develop on the ripening tomato fruits, making them inedible. Potato tubers that looked fine at harvest can rot in storage. Though I'd never had late blight in all my years of farming, I was aware that in wet and cool conditions it's a disease that can spread like wildfire. And wet and cool conditions were exactly what we'd had for an entire month.

I also knew that late blight, known among the cognoscenti as *Phytothphora infestans*, was in the air. I'd heard reports that the

I'd never had late blight in all my years of farming.

dreaded Irish Potato Famine disease had been sighted on Long island and in other parts of New York State, but somehow I'd never imagined it would find me. It was presumed to have come into our region on infected seedlings distributed by Bonnie Plants, a huge supplier of plants, with headquarters in Alabama and greenhouse complexes in several states. The infected seedlings had shown up in many of the big box stores in the eastern United States—Walmart, Lowes, Home Depot. Even after being told that many of them carried a deadly and highly contagious disease, these money-gobbling colossi were rushing to sell their inventory of plants to home gardeners before it was too late. The worse the plants looked, it seems, the deeper the discount customers were offered. (In case you're getting nervous, late blight is strictly a disease of plants. It does not in any way affect humans.)

I could tell that Maire did not enjoy being the bearer of such bad news. In an attempt to soften the blow she went on to say: "It's not just you, Keith. Late blight is showing up everywhere. I bought some plants in Lowes yesterday, to take back to our lab for testing. You wouldn't believe how sick they looked. I almost got into a fight telling one woman not to buy a large potted tomato that was reduced from $20.00 to $2.50. It clearly had the blight, but the bargain price was too much for her to resist."

The Irish potato famine was no joke. Potatoes, once introduced to Ireland, soon became the staple crop of the Irish peasantry. They grew well and harvests were bountiful, until late blight came along. Over the course of several years, starting in 1845, the disease annihilated the potato crop, resulting in the death by starvation and related causes of roughly a million people. And it led to an enormous surge in immigration to North America. The problem was exacerbated by farmers doing what they had always done: saving seed to plant the next year. The infected tubers, when replanted, just ensured another round of disease. Late blight is indigenous to Central and South America, the home of potatoes. The outbreak in Ireland was probably caused by infected tubers coming up from Mexico, crossing the United States, then making their way to Ireland, by boat, from New York. In the cool, damp Irish weather, the disease exploded.

"Don't plant any of these spuds next year" was Maire's first utterance of advice soon after we had moved into a field of potatoes. "Even if they look okay. I hate to say it, but you've got the beginnings of blight here too." She suggested that I spray all the tomatoes and potatoes with a copper-based fungicide as soon as possible. I groaned, knowing from very recent experience how onerous a task this would be. She then added that, if I didn't want to spray, I might try going to church on Sunday and putting twenty dollars in the collection box. I couldn't tell whether she was serious about this. But, since I seldom find myself in a church, except for the occasional wedding or funeral, it didn't seem like much of a plan. Which left me with copper.

As Maire's car disappeared up the driveway, I got on the phone to see what I could find. It was the afternoon of July 3rd, and most companies selling agricultural pesticides were closed. I did, however, manage to place an order for a twenty-pound bag of something called NuCop 50, one of the few copper fungicides permitted under organic standards. But I wouldn't be able to get hold of it until Monday, at the earliest. There was nothing more I could do.

I tried, with little success, to expunge all thoughts of blight from my mind, and went on to harvest mint, snow peas, and wild mulberries for the Saturday market. With all the rain we'd been getting, at least there was a veritable forest of good-looking mint to cut from.

As expected, the July 4th market at Union Square was slow. That meant there was plenty of time to chat with neighbors and customers. Early on, a fellow organic farmer and friend, Zaid Hadieh, stopped by to do just that. When I told him that we had late blight on the farm he threw up his hands and said, "You've got to spray with copper. Now." I told him I couldn't get any 'til Monday. He said that might be too late and described the devastating losses he had suffered due to the disease some years before. Generous man that he is, Zaid offered to give me a couple of pounds of copper fungicide if I could find someone to make the trip to his farm to pick it up. A few phone calls later, an arrangement was made. One of my workers, Josh, was willing, on his day off, to drive two and a half hours north to obtain an emergency supply of copper. The next day we started spraying.

We're now in the third week of July and the weather is stabilizing. It's still surprisingly cool for mid-summer, but the rain has eased off a bit. Over the past two weeks, I've become a regular nozzle-head, spending about a quarter of my time either mixing fungicides, riding on a spray tractor or, worst of all, walking through the fields with a long hose and spray gun. I don't know what good it's doing. The Downy Mildew has moved from its place of origin to a more distant patch of onions. The stressed plants are at least beginning to form little bulbs and it looks like we might have something to harvest. A couple of small plantings of scallions bit the dust pretty hard, but so far, our garlic seems to have resisted the disease. For this I am very grateful.

Latest news reports warn that late blight is now well established up and down the eastern states and has moved inland as far as West Virginia and Ohio. It has definitely spread through our first planting of tomatoes, despite the application of copper. It's just a matter of time before more plants end up in the big hole.

This year, we planted a total of 2,300 tomato seedlings in three different locations, a few weeks apart. All of them were started in our own greenhouse. The second planting is looking a little better. It has patches of blight but some plants are not yet showing any sign of disease and seem to be growing well enough. Maybe the copper is doing its job. I'm sure the drier weather has worked in our favor but worry what will happen when it starts to rain again. It's too soon to predict how the third planting will turn out.

About a third of our potatoes now have some degree of late blight, but the combination of dry weather and copper seems to be holding the disease in check. The plants are large and so have good reserves of stored energy to form their tubers. If we can just keep them alive for another two or three weeks, I think we'll have

Our garlic seems to have resisted the disease. For this I am very grateful.

potatoes to sell. Last night, a farmer friend and fellow garlic grower from the north country told me that I should consider mowing all the aboveground growth off the potatoes within a week or two because the late-blight pathogen needs living tissue to survive. He said that if I let the tops dry out or decompose and then wait a couple of weeks before digging, there's a fair chance that most of the tubers underground will be fine, though perhaps smaller than usual. I hope he's right.

We ended our conversation on a hopeful note, reflecting that so much adversity might lead to greater empathy for others, especially other farmers, many of whom are in the same boat I'm in, or worse. And, who knows, it might even lead to a little personal character building. That would be a nice silver lining.

Rocambole Garlic

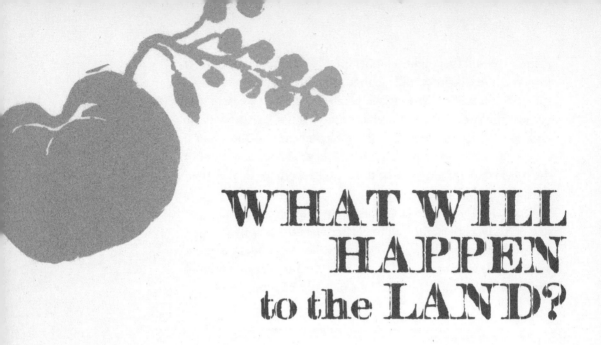

WHAT WILL HAPPEN to the LAND?

We paid $215,000 for our farm in 1986. At the time it seemed like a lot of money. And it was. We cashed in our mutual funds and took out a hefty mortgage. What we got in return was 88.5 acres of variably wooded and open land, a small, somewhat run-down farmhouse with a long history, a reasonably good barn, and a couple of ramshackle sheds. Previous owners, and there were several in the half century before us, and an unknown number before that—all, it seems, engaged in dairy farming—probably lived here happily enough, but, judging from the appearance of the place when we moved in, they did not prosper financially.

Over the years we've made several improvements to the house and outbuildings. The place looks better than it did when we arrived, but it's still the same piece of land, with vegetables on it now instead of cows (unless you count Tiny Tim and his bovine harem during the summer months). There is, however, one big difference: Today, the farm is worth a lot more money. If we put it on the open market, it would probably fetch four times what we paid for it, possibly more.

On one level this large appreciation in value is very pleasing. By any measure of capital growth or investment, we've done rather

Field and Farm

Downy
Woodpecker

well. My father and mother, now both gone, would have been proud of me. "Keith's got his head screwed on all right, after all," I can hear my father saying. In my less responsible and more itinerant youth, there were definitely periods when they considered the secure attachment of that part of me above the shoulders to be in doubt, and harbored perhaps legitimate parental concerns regarding my ability to prosper in the world. Yes, they would be pleased and proud and a little relieved, and perhaps I can share in those sentiments. There's just one problem. In order for us to receive four times what we paid for our farm, to cash in, so to speak, we would have to sell to a developer, who would, in short order, divide the land into building lots, put in a road, and erect some fifteen or twenty houses.

The unavoidable truth is that no young would-be farmer, unless endowed with a trust fund or some independent income stream, could afford to buy our place and try his or her hand at vegetable growing as I did back in 1986. There's just not enough money in farming to justify a start-up cost of that magnitude. Not only would the land be prohibitively expensive, there would follow many other costs, such as the purchase of equipment and machinery, seeds, plants, labor, insurance, property taxes, and so on. And, as any small-business owner knows, the chance of turning a profit in the first year or two would be slight. It pretty much boils down to one thing: Sale of the farm on the open market means the end of the farm.

Over time, a piece of land can grow on you. Both my wife and I have become fond of this place we call home, more fond than we might have expected. It holds the history and memories of some large part of our lives. And it seems to have a life of its own—to be, in many ways, so much bigger and richer and more complete than we, as individuals, are. An extraordinary number of animals, birds, fish, insects, and plants also regard the place as home and have done so since long before we arrived. Many, including the raccoons and opossums who dine on mulberries in the summer and the woodpeckers whose rat-a-tat-tats echo through the woods, and the wild turkeys who forage along the edges of the fields, are year-round residents;

others, like the barn swallows who raise their young in the barn or the green heron who frequents the marshy edges of the pond, stop by for spring and summer as they pursue the migratory cycles of their lives. To sell the land to a developer would be to disenfranchise and displace a great many of these nonhuman inhabitants.

There are also the vegetables. For nineteen years we've been growing produce and trucking it down to New York City two or three times a week. We have several hundred customers who look forward to our return to market each year and who place real value on the fresh, organic vegetables and herbs we sell them. If the farm becomes a housing subdivision, they, too, will be disenfranchised. They will have lost one more source of local food and will be left to rely on the whims and maneuvers of a faceless, industrial food system and the uninspired offerings of local supermarkets. Their experience of eating will be less personal, less connected to place, and probably less healthy and secure. When the only garlic in town is shipped from China and the only tomatoes they can find come from a corporate farm in Mexico, they will have entered a global, homogenized, and more insipid world of food.

Each of these losses—the loss of history, the loss of habitat, the loss of a local, sustainable, and verifiable food system—would not be recognized on any balance sheet or reflected in our nation's economic trajectory. Yet the losses would be real; for some people and for some animals, they would be profound.

In the relatively short history of the European conquest and appropriation of North America and the spread of our much-vaunted market economy across the land, if not the entire planet, the potential disappearance of one small farm is a minuscule affair, not worthy of mention. But if you multiply that loss by ten thousand, it perhaps does become significant. For it is not just one farm that I am talking about. The potential fate of our farm is but a single example of a relentless trend that is changing the landscape we live in and our proximity to fresh food. This trend can be summed up in less than a dozen words: The economics of farming cannot compete with the economics of development.

Green
Heron

Almost every farm within seventy-five, or perhaps a hundred, miles of New York City is under intense development pressure—to be sure, some more so than others, but it is the rare farm for which the clock is not ticking. Similar radii might be drawn around many of our other great cities, for this is a national trend. In the United States we lose more farmland every year than any other Western country does. Much of this land is lost to a lavalike flow of sprawl around our largest cities.

Most of us know that family farms have been disappearing from the landscape for a long time, but few realize the extent to which this trend has accelerated in the last several years—spurred on, no doubt, by low interest rates and the trauma of 9/11. With great haste and little reflection, we are moving toward a time when there will be, at most, a mere handful of small farms to serve our major cities. These will be showpiece or token farms, mostly operated by liberal-leaning colleges and well-endowed nonprofits. They will not be able to meet the growing demand. It is ironic that the popularity of small farms and local food production should reach such exalted levels just as so many of the former, and by extension so much of the latter, slip from reach.

I came to farming in middle age. I've put in more than twenty years. They have been the most challenging, the most rewarding, and the most dynamic years of my life. I am profoundly grateful to have had them. But now I am getting a little tired; my body does not work as well as it used to. It has more aches and pains and they are more persistent. The ninety-degree days of summer leave me totally exhausted, unable to do much more than eat and sleep. A cold, wet day of harvesting in November makes my joints ache and causes the arthritis in my fingers to flare up. Each $2 bunch of collard greens or Red Russian kale that I cut and wrap a rubber band around exacts its own small toll.

My patience for training new employees each year is not what it used to be. To get the job done—to get my young, American crews to work productively and efficiently—is a constant and sometimes wearying challenge. It requires that I be present and work alongside them much of the time, that I set a good example and a good pace. Ten years ago I could do this without much trouble. Now it is not so easy. Moreover, keeping these young men and women content with each other and with me, both in the field and in the domestic realm, often seems like an almost unattainable goal, no doubt made more so by my own shortcomings.

Other farmers tell me to hire Mexican migrants who are experienced in agricultural work and, they insist, more productive than most young Americans. This is almost certainly true, but I don't speak Spanish and I feel it's a bit late to teach an old dog (this one) new tricks. And the truth is, though from time to time I may feel some frustration with the pace and progress of a day in the fields, in the end I enjoy my young indigenous workers and their youthful energy. And, perhaps, they keep a certain energy alive in me.

For most who come here, one season is enough to satisfy whatever idealistic or pastoral itch they might have had but, when they leave, they are fitter and stronger and more physically able than when they arrived, and they have an understanding of what it takes to grow food and work outside in all kinds of weather. And there are those (perhaps one in ten) who really take to farming, who have a natural gift for it. You can see it in their bodies, the spring in their step, and the light in their eyes. It is a pleasure and delight to work with them and, when they leave, perhaps to farm elsewhere, it is gratifying to think that I, or this piece of land, might have left an impression that will stay with them and lead them on to a farming life.

For the past several years I've been on the lookout for a good comanager—a person who will assume some responsibility for the daily running of the farm and share in the profit; someone I like and who likes me; someone I can get along with; someone who knows how to get work done and who can train and motivate others to do the same; and someone who will stay with the farm for the long haul. I have had a few managers over the years, but no one has lasted more than a season or two and none has been quite up to the job. Mostly, they have been good workers but have not

They will have lost one more source of local food and will be left to rely on the whims and maneuvers of a faceless, industrial food system.

excelled as managers, have not been able to interact effectively with the rest of the crew. It is not an easy job. When I talk with other aging farmers about my desire to find a good manager, they invariably laugh and say: "Isn't that what we all want?" They too are feeling the wearing effects of time and in many instances have no willing heirs—at least to the work of farming the land. What they do have, though, in most cases, is a piece of real estate that will finance their retirement.

My plan is to keep living on this farm until I am no longer able to—perhaps ten more years, perhaps twenty, whatever portion destiny allots me. But I doubt that I will last much longer as the sole driving force of a productive vegetable operation. Someone else will have to do that, or at least share that role with me, or the land will lie fallow, or go to some less intensive agricultural use—either of which, despite the loss of choice to my customers in New York City, might not be such a bad thing.

But in the end, the big question remains: What will happen to the land when my wife and I are ailing, or no longer able to live here, or simply gone? Sale to the highest bidder is one possibility. But I hope not. Both of us want to avoid this because we know what it means, but we may find ourselves in ill health or financial straits that leave little choice.

Donation of the development rights to a land trust or some other conservation organization is a second option. This would enable us to continue living on the farm and at some point sell it with a conservation easement attached to the deed, requiring the purchaser, and all future buyers, to maintain the land as open space. In legalese, such deed restrictions are said to be effective "in perpetuity." But after the 2005 U.S. Supreme Court decision allowing private property to be taken in New London, Connecticut, via eminent domain, for what the Court deemed to be a higher, private use (namely, a more lucrative one), the donor of a conservation easement must wonder just how long "in perpetuity" might be.

Donation of the development rights would give us some tax advantages for a few years but would not, by a long shot, result in the fourfold profit that an unencumbered sale of the farm on the open market would probably bring.

A third possibility is the sale (rather than donation) of development rights to a land trust or conservation organization that would be similarly pledged and legally bound to ensure that the land remain open space, preferably in sustainable agricultural use. Assurance, at least for the foreseeable future, that the land remain open space is achievable, but there is no way to insist that future owners grow vegetables or milk cows or engage in some commercial farming enterprise. Of course, while the land is open, this remains a good possibility. If the farm becomes a housing development, agriculture, in any of its forms—this side of another Ice Age—is unlikely in the extreme.

Naturally, selling the development rights is more attractive to my wife and me than outright donation. We would receive less money than if we were to sell to a developer but more than if we simply gave the development rights away—maybe some amount roughly in between. That seems like a better deal—one that perhaps my father would approve of. Both we and society would share in the cost of preserving a farm, and both parties would reap the many tangible and intangible benefits that would accrue over time.

Unfortunately, most land trusts, which are dependent on private donations and staffed largely by volunteers, do not have the funds available to purchase development rights or conservation easements. Even when development rights are freely donated, land trusts will often charge the donor a custodial fee of several thousand dollars, which is used to monitor the property and ensure that it remains undeveloped.

Neither my wife nor I are in desperate need of money. We are both in reasonable health. We do not have children to send to college and we are not big spenders. As things stand today, our farm is not in immediate peril. Many other farmers are less fortunate in this regard, or have greater needs and obligations. I have no doubt that most farmers would prefer to see the land they have labored on over many years, and perhaps grown to love, remain in agriculture. But for most, the cost of preserving that land, without assistance from society, will be too great; the promise of financial independence through sale on the open market will, understandably, be too attractive.

The questions then arise: Does the rest of society care? Does the rest of society value having small farms in its midst, having local farm stands and food available, seeing a field of corn, or hay, or vegetables, or a tractor on the road from time to time? Do cities like New York and Boston and Philadelphia and Washington, D.C., truly value the fresh food that is brought to them through farmers' markets and urban CSAs? Is the relationship between farmer and consumer, rural and city folk, worth preserving? Do local and regional food sources enhance our lives and afford us a certain security in an age of open-ended conflict and skyrocketing fuel costs that may undermine the global traffic in food? If the answers to these questions are yes, yes, and yes, then one must ask, do we care enough to do something about it? That is the real question.

On the radio the other day, I heard that Orange, the county in which we live, is the fastest-growing of New York's fifty-seven counties. This statistic might excite the local chambers of commerce and economic development people. It is probably good news for builders, masons, plumbers, and electricians. It means bigger tax budgets for towns and municipalities. It means more roads, more malls, more houses, more schools, more services. There's just no getting away from it. This is the path we are on. But, there's a stubborn side of me that keeps insisting that growth does not have to mean removal of all farms—there's no economic law requiring that. If we came together and developed a plan that rewarded farmers for preserving their land, we could have houses and farms side by side. Certainly, it would require some investment of money and will, but we and future generations would end up the richer for it.

There's a stubborn side of me that keeps insisting that growth does not have to mean removal of all farms.

A FARM in PERPETUITY

I n *December of* 2007, my wife and I signed a conservation easement formally waiving the right to develop our farm. It was the
culmination of a long process, the goals of which were always to
protect the land from future development and receive some compensation for doing so.

On the day of the signing, we received a check for $235,000,
which was exactly half the assessed development value of the property. The other half was viewed as a donation, by us. Under the terms
of the agreement, we retain full ownership of the farm and are able
to sell it, should we wish to, but neither we nor any future owner can

ever draw up a subdivision plan and grow a crop of houses. This means that the property's resale value is considerably diminished. It also means that it should be more affordable to some future farmer.

The process began eight years earlier when we filled out our first application to the State of New York's Farmland Preservation Program, offering to sell our development rights for a bargain price. Presumably because of the large number of farms applying and insufficient state funding, we were not accepted into that program.

Then a few years ago, we heard about legislation to preserve farmland in Orange County, financed by the Orange County Open Space Fund. It seemed like our chances might be better with a program focused on our own county. We reasoned that the huge loss of farmland in Orange over the previous decade might work to our advantage—there couldn't be that many farms left to protect. So we rolled up our sleeves and went to work on another rather daunting application—with maps, photos, and other attachments, it ended up being over seventy pages long, and fifteen copies were needed.

Our first application to the County met with rejection. No big deal; we were used to that. We tried again the following year and, after a wait of several months, received a letter from the Orange County Department of Planning and Development notifying us that our farm was "conceptually" approved for protection. That was very encouraging, but there was still a distance to travel.

◆

I've always been fond of land, especially land in a somewhat natural state. Parking lots, malls, subdivisions, and golf courses just don't turn me on. That's not to say the tillable fields on our farm closely resemble what nature would design. The long rows of vegetables and herbs simply wouldn't be there without the industrious efforts of my workers and me. But vegetables don't swallow up the land. They just borrow it from one season to the next, and give it back during the winter. Moreover, they generate income, pay the bills, and employ people. And, of course, they feed people too.

Though I may often regard my vegetable fields with a certain custodial pride, to me they are a less interesting part of our farm. If I just stop to look and listen, there is so much more: the tangled hedgerows along the edges of the fields, the ponds that have become home to numerous aquatic creatures, the perimeter zones of mixed hardwood forest with patches of hemlock and pine, the rocky ridge on the southern border, and the crooked creek that runs through the lowland pasture. To me, all this is the wild heart of the farm. Even in its quietude, the land pulsates with life—an almost incalculable number and variety of plants, animals, birds, insects, microbes, all existing together in a rough, give-and-take sort of harmony. A housing development would obliterate this.

To me, all this is the wild heart of the farm. Even in its quietude, the land pulsates with life.

Killdeer

So long as land is just another commodity and the goal of achieving maximum profit on investment rules the day, then houses will consume farms every day of the week. Often, the land is used with terrible inefficiency. Picture a 3,500 square foot house in the middle of a 100,000 square foot lawn. That's very common. Just the thought of mowing all that grass is enough to give me a headache. But what is really troubling is the permanent withdrawal of so much fertile land from agriculture. It's a no-brainer. If you want local food, you need local farms. The bumper sticker on my truck puts it more succinctly: "No farms. No food."

◆

You might think that when we heard from the county's planning department that our offer to sell the development rights on our farm was "conceptually" approved, it meant most of the hard work was behind us. In fact, that encouraging letter just marked the beginning of the next stage of the process. There was still plenty left to do and numerous parties whose services would be needed. For starters, we were advised to retain a lawyer, which we did.

A full survey of the property was required. This took some time to set up and then a couple of weeks to execute. Next, a professional real estate appraisal was needed. The appraisal ended up being an almost ninety-page report which included numerous maps, photos, and physical descriptions of the farm and surrounding area, as well as data on comparable land sales both in the vicinity of our farm and elsewhere in the county. The appraiser arrived at the following conclusions regarding the value of our land:

"Before easement" value of land:	$885,000.00
"After easement" value of land:	−$415,000.00
Value of conservation easement:	$470,000.00

In our initial application we had agreed to accept 50 percent of the development value of our property. This is why our check from the county was in the amount of $235,000. Before the final deal was struck, we settled on other requirements, including:

A title search on the property;

An inspection and evaluation by an agent of the Orange County Soil and Water Conservation District;

Agreement to a Stewardship Plan for the land;

A thorough Baseline Study (documenting, with photographs and maps, the exact condition of the farm and its various structures immediately prior to the signing of the easement). This would enable future monitors to detect if changes were made that violated the terms of the easement.

The conservation easement that we eventually signed is a legal agreement between three parties: my wife and I, as the owners of the property; the county of Orange, as the purchaser of the development rights; and the Orange County Land Trust, a local, nonprofit organization with the mission of protecting open space. The Land Trust would end up holding the easement and be responsible for ensuring that its terms are kept and that no disallowed uses occur on the property. The Land Trust also helped to coordinate the survey, the appraisal, and the baseline study.

The easement itself is a dense, 22-page legal document. It defines the ground rules and divides the property into three zones:

1. The Farmstead Complex, where the existing buildings are and where certain additional buildings to support the farm operation may be constructed.
2. The Farm Area, which comprises the majority of the property and is where most of the farming will take place.
3. The Resource Protection Area, which is essentially a twenty-foot buffer zone on either side of the stream that runs through some of the farm's woodland and pasture.

The easement then goes on to spell out exactly what can and cannot be done in each area. There is a lot of detail. For example, some logging can be done in the Farm Area, but not in the Resource Protection Area. Within the Farmstead Complex, one additional residence can be constructed, to support the farm oper-ation, but we are not allowed to build an inground swimming pool or a heliport (that put the kibosh on my lifelong desire to com-mute to work in a chopper!). In the Resource Protection Area, no more than fifteen animals (presumably cows or horses) can be pastured at one time, a restriction designed to minimize stream bank erosion.

Most of the terms of the easement make good sense, even the ban on swimming pools and heliports. The County and the Land Trust would prefer the farm to remain in productive agriculture rather than become the weekend playground of a bank manager looking for something to do with his or her latest bonus, or possibly a hedge fund operator who would surely feel entitled to install a private swimming pool and helipad after playing fast and loose with his clients' money.

Until the actual day of the signing, or close to it, the conditions of the easement were not set in stone. The Land Trust and the County wanted more restrictions on the property and we wanted fewer. They were spending public money and, understandably, wanted to get as much for that money as possible. My wife and I, on the other hand, were keen to retain some flexibility. Though our wish to make future changes may have been limited, we understood that the resale value of the farm would diminish as more conditions were placed on how the land could be used.

Sometimes this put us in a somewhat adversarial position with respect to the Land Trust and the County. For the most part, our differences centered on small details (like the number of cows permitted in the pasture, or the width of the buffer zone along the creek) and most of them got worked out to all parties' satisfaction. The possible future paving of the driveway was one issue, though, where we had a real difference of opinion.

Our driveway is long and steep and vulnerable to heavy rains and snow melt; pot holes and ruts soon develop regardless of how much stone you put down. We've got friends who are loathe to subject their vehicles to its sometimes punishing and unpredictable surface. The Land Trust wanted the driveway to remain gravel or some other loose material. We wanted the option to pave the driveway at some point in the future. Here, we held firm and prevailed, on the condition that we use materials and methods that would be least disruptive of the natural topography. No problem there.

◆

A week prior to the signing of our easement, the Orange County Legislature held a public hearing at which anyone could speak in favor of, or against, the use of public funds to protect our farm. We were strongly encouraged to attend and make a statement. When the time came, I got up, went to the podium and, somewhat nervously, delivered a short written speech in support of farmland preservation. No opinions to the contrary were expressed. Soon after, the legislature voted to approve.

Finally the day came. The various lawyers assembled in a conference room (ours showed up a little late), along with representatives of the Land Trust and the County, and, of course, my wife and I. Numerous papers were signed, a few jokes were made. We were presented with a large check and congratulated on taking the step to preserve our farm "in perpetuity," to use the language of the easement.

*Fresh snow
had fallen
during the
night. There
was still a
coating of ice
on the trees.*

As we drove away, I reflected on that term, "in perpetuity." It is, of course, a very long time. Perhaps "as long as possible" would be closer to the truth. So much can change. Just the other morning I was listening to the BBC World News. After covering the mounting civilian casualties in one of our planet's numerous "small" wars, the announcer introduced his next segment with the words: "And now for some really bad news." My ears pricked up as I learned that our galaxy, the Milky Way, is on a collision course with the Andromeda Galaxy. I did not know this. Even more troubling was the revelation that scientists now believe the collision will occur much sooner than had been expected. Admittedly, though, the big event is not scheduled to occur for many millions of years. But, when it does happen, everything we know will disappear, even farms with conservation easements. For a moment, it made me wonder if all our efforts were really worth it.

Then I stepped outside and breathed in the clear winter air. Fresh snow had fallen during the night. There was still a coating of ice on the trees from the day before. It glistened in the early sun. A few crows had assembled on the upper branches of a dead elm tree at the bottom of the driveway. They seemed to be conversing about something. I listened to them for a minute or two and then experienced a small moment of epiphany. Perhaps it was the language of the crows. I don't know. But I understood in that moment that our efforts were worth it, even if nothing is ever really "in perpetuity."

EPILOGUE

To those of us for whom there is no other home than earth, however imperfectly we might inhabit this temporal world, a piece of land can assume the nature and aspect of a lover, though she be but a small piece of a great whole. Stray from her though we may, we always return to her breast to quench our desire and replenish our spirit. It is in the land's boundless profusion and regenerative, forgiving ways that we find solace and a kind of wholeness.

I never would have guessed that I would end up being a farmer. As a youth I showed no interest in my parents' well-kept garden—neither the vegetables that mostly my father grew, as befit his more practical character, nor the shrubs and flowers that my mother, who was more artistic in nature, devoted many hours to. If my father insisted, I may have consented to mow the lawn on occasion, or clip the hedges between our house and the neighbors', but I was seldom seen with a rake or shovel or packet of seeds in my hand. The garden was too tame a place for me. I always looked to the wild mountains and to adventure.

After a life of travel and travail, Voltaire's formerly optimistic hero, Candide, has learned that the world of men is too random, cruel, and unjust for his taste. He elects to turn his back on the larger human community. All that remains for him is to cultivate his garden in the company of a few loyal companions. In this he finds a certain comfort and content.

I do not mean to elevate myself to the level of Candide, nor imply that whatever minor hardships I might have endured in my life are in any way comparable to the grotesque and cruel treatments meted out to him and those close to him. Indeed, this is not the case. For the most part I have been the beneficiary of good fortune, generosity, and love. I should have no grounds for discontent.

But, for whatever reasons—temperamental, philosophical, or perhaps some misalignment of heavenly bodies at my birth—I must admit to a certain recurring disenchantment with the human enterprise, especially as it manifests itself in the various concentrations of power and greed that shape our lives. I fear that we have got ahead of ourselves somehow, that we have too much hubris and destructive might for our own well-being and the well-being of the planet, and that ultimately we are at grave risk of running amok on a very large scale. I fear for my fellow man but my sympathies also lean, increasingly, toward the rest of nature, from which we humans seem eager to divorce ourselves and abrogate all sense of stewardship, even as we overwhelm, exploit, and abuse it. What kind of god has taught us to do this?

In his classic book, *A Sand County Almanac*, published in 1949, the great conservationist Aldo Leopold wrote of the need for a land ethic, a new way of thinking about land that reaches beyond our immediate economic and utilitarian interests. Leopold saw the land as a community of interdependent parts—soil, water, plants, animals, with man as a member of the community, not its master. He understood that an assault on any element of the land community would affect the health of the whole. He argued, with great eloquence, for an ecological consciousness and a gentle touch.

A small farm is, I believe, a place where one can work to develop an ecological consciousness and live in some measure of harmony with one's surroundings, without withdrawing from the rest of humanity to the extent that Candide did. I believe it is possible to both cultivate one's garden and remain engaged in human society. Growing vegetables for market requires that I keep one foot grounded in the practical and hardscrabble world of commerce and competition. But, so long as I can keep my balance, the other foot will stay firmly planted on the good earth from which I draw sustenance, inspiration, and instruction.

For these reasons and in these times, I commend the farming life, even if it is only a gesture.

APPENDIX

KEITH'S FARM BIRD LIST
May 1–November 20, 1995
Compiled by Rob Morrow

Northern mockingbird
Song sparrow
Robin
Cardinal
Blue jay
Brown-headed cowbird
Red-winged blackbird
Rose-breasted grosbeak
Eastern bluebird
American goldfinch
Canada goose
Turkey vulture
Red-tailed hawk
Barn swallow
Tree swallow
Catbird
Tufted titmouse
Scarlet tanager

Common crow
Rock dove
Starling
White-throated sparrow
House finch
House sparrow
Osprey
Mallard
Wood duck
Blue-gray gnatcatcher
Common yellowthroat
Yellow-rumped warbler
Downy woodpecker
Northern flicker
Eastern phoebe
Northern oriole
Black-and-white warbler
Black-capped chickadee

Yellow-Billed Cuckoo

Rufous-sided towhee
Eastern kingbird
Blackburnian warbler
Great blue heron
Veery
White-breasted nuthatch
White-crowned sparrow
House wren
Chipping sparrow
Field sparrow
Red-eyed vireo
Blue-winged warbler
American redstart
Yellow warbler
Northern waterthrush
Barred owl
Great crested flycatcher
Mourning dove
Common grackle

Indigo bunting
Eastern wood pewee
Green heron
Yellow-billed cuckoo
Ovenbird
Cedar waxwing
Belted kingfisher
Great horned owl
Junco
Northern harrier
Ruby-crowned kinglet
Hairy woodpecker
Wood thrush
Sharp-shinned hawk
Kestrel
Hermit thrush
Ring-necked pheasant
Purple finch

ACKNOWLEDGMENTS

My *thanks go* out to the many faces—human and nonhuman alike—who have shown me friendship and love and shaped me along the way. There is not space here to recognize all of you. Some, however, warrant special mention—my parents, Stan and Pauline Stewart, for providing the bedrock of my being; my sisters, Claire and Barbara, who read me stories at night; my wife, Flavia Bacarella, who has shared this farming journey with me, taught me how to love, and so touchingly illustrated our book; my mother-in-law, Angie Bacarella, who always found me the best clothes and who loved working on the farm.

My sincere thanks also to Jerry Novesky and Janet Crawshaw of *The Valley Table* magazine for encouraging the writer in me and taking a chance; and to Matthew Lore of The Experiment for doing the same. Thanks also to: Andy and Ida Burigo, Sally Schneider, Leslie McEachern, Ingimundur Kjarval, Geoff Bresnick, David Church, Wayne Decker, Peter Hoffman of Savoy, David Stern of the Garlic Seed Foundation, Maire Ullrich of CCE, Jim Crawford of New Morning Farm, fellow farmer/writer Tim Stark, and my neighbor, friend, and tractor advisor Ted Stephens.

Thanks to many old friends, some no longer with us, and the memories they have given: Garth Melville, David McNicoll, Michael Guiniven, Burton Silver, Alan Radcliff, Robert Cable, Denys Whalley, Paul Fuller, GeorgAnn Stewart-Stand, Peter Eckman, Hansford Rowe, Bill Norris, Carolyn Bell, Alice Morris, Gerhard Riethmuller, Eddy Bennett, Tom and Jane Hatley, Bob Aldridge, Steve Smith, David Petraglia, Libba Marrian, Guy Jones, Andy Van Glad, Jim Miedema, Don Lewis, Dave Distler, Neil Ostergren, Mary and Gary Kerstanski, Gary and Mary Ellen Calta, and the infamous and inspirational Yale School of Forestry & Environmental Studies, Class of 1980—every single one of you.

Thanks also to my many hundreds of wonderful customers at the Union Square Greenmarket in Manhattan—you who have given support and encouragement to my farming endeavors, always with such warmth, humor, and cash. Without each of you, our small farm would not be the place that it is. And thanks to the sterling management and staff, past and present, at Greenmarket and the Council on the Environment of New York City (in particular, Michael Hurwitz, Marcel Van Ooyen, Lys McLaughlin) and Greenmarket's founding fathers, Barry Benepe and Bob Lewis, for your stewardship of a true win-win program for farmers and city dwellers alike.

Finally, I thank my young and sometimes not-so-young American workers—so many names and faces over more than twenty years—each of you, whose sweat, occasionally blood, and perhaps even tears have enriched this rocky soil, each of you who has labored alongside me in all manner of weather, through good times and bad, to bring decent, honest vegetables and herbs to our city neighbors. In the approximate order of your appearance on the farm, I thank you all:

1989 ❧ Mitchell Stern, Lucrecia Crimmins, Chuck Crimmins

1990 ❧ Julia Paulsen, Emmett Mullin

1991 ❧ Sharon Yablon, Michael Grossi, Peter Wilshusen, Paul Amhrein

1992 ❧ Kristin Speegle, Sharon Yablon, Daniel James Mac-Neil, David Strasfeld, Michael Grossi

1993 ❧ Michael Todd Juzwiak, Eva Barr, Larisa Vermeulen, Michael Grossi, Alan Page, Mitchell Stern, Kristin Speegle, Thomas Little, James Garzelloni

1994 ❧ Alan Page, Lisa Lindgren, David Bechtel, Ivan Siff, Michael Moon

1995 ❧ Christianne Corbett, Lori Scarpa, Christopher Wall, Robert Morrow, Conroy Smith, Julanne Tyler, Christopher Conrad, Eric Hinson, Emily Bayley

1996 ❧ Christopher Amato, Lisa Amato, Timothy Bartlett, Kimberly Morrell, Ehren Reynolds, Kirstin Reynolds, Carrie Spring, Moneque Williams

1997 ❧ Jason Grzanna, Heather Henry, Bill Freese, Matthew Vivrett, Sandra Cavalieri, Emily Shaw, Patrick Martin

1998 ❧ Graham Hawks, Laura Moul, Matthew Vivrett, Ione Lindroth, Jessie Burgoyne, Sandra Cavalieri, Luca Gasperi, Paul Humm

1999 ❧ Kevin J. Toomey, Martha Gross, Stacy Dorris, James Breneman, Jennifer Fayocavitz, William Irons, Richard Trichon, Jason Grzanna, Aaron Seward

2000 ❧ Caroline Adams, Ben Bechtold, Sarah Clifton, Jason Grzanna, Jim Poe

2001 ❧ Eva Brisker, Kerri Smith, Kate Canney, Carol Landry, Yvette Garcia, Ken Burns

2002 ᴥ Eva Brisker, Yvette Garcia, Dan Zabara, Jacquelin Goveas, Alison Masters, Eric Anderson, Rocco, Samantha Dinar, David Pesnichak

2003 ᴥ John Seaver, Jessica Napper, Jessie Burgoyne, Charles Andy Wingard, Kristin Van Fleet, Jacob Diaz, Jenny Kessler

2004 ᴥ Jason Grzanna, Eric Anderson, Shoshana Woodworth, Theresa Mycek, Barbara Berrang, Robert Brian Autry, Jess Dolan

2005 ᴥ Stephanie English, Daron Hoffman, Melanie Medeiros, Monica English, Russell Ben Lewis, Landon Jefferies, Emily Myers, Lorri Myers, Peter McAvoy, John Franklin Egan, Ann Novak

2006 ᴥ Russell Ben Lewis, Landon Jefferies, Monica English, John Franklin Egan, Nicole Sinovsky, Nicole Roy, Ben MacDonald, Lisa Weiss, Lila Matsumoto, Vanessa Leach

2007 ᴥ Ben MacDonald, Monica English, Amanda Andrews, Melissa Weinberg, Adam Weg, Devin Walker, Alex Cuff, Ezra Bassel, Julia Landau, Damon Lewis, Stephanie Lewis, Jamie Chiarello, Colin MacDonald, Alex Enarson-Herring

2008 ᴥ Amanda Andrews, Monica English, Dan Criss, Rachel Hestrin, Coco Roy, Joe Piekarski, Graham MacDonald, Dylan Maloney, Frederico Gimenez, Christopher Long

2009 ᴥ Lisa Gaeddert, Juliette Enfield, Daniel Berry, Joshua Passe, Matthew Ready, Qayyum Johnson, Sarajane Snyder

2010 ᴥ Joshua Passe, Daniel Berry, Matthew Ready, Lisa Gaeddert, M. Weatherly Thomas, Patrick Kelly, Ben MacDonald, Kysa Heinitz, Alex Cuff

ABOUT the AUTHORS

For twenty-four years, KEITH STEWART has been the proprietor of Keith's Farm, in Orange County, New York, where he grows, with the assistance of six or seven seasonal workers and under certified organic conditions, one hundred varieties of vegetables and herbs and maintains a small grove of fruit trees. Stewart is one of the longer-standing purveyors at New York City's Union Square Greenmarket, where his stand has a devoted following that includes many restaurateurs and food writers, among thousands of others.

Stewart has appeared on numerous TV and radio shows, including the Food Channel's *Follow that Food*, PBS's *Chefs A' Field*, the *Leonard Lopate Show*, and has been featured in publications including the *New York Times* and *Gourmet*; for more than ten years his writing has attracted legions of fans in *The Valley Table*, the Hudson Valley's only magazine devoted to regional farms, food, and cuisine.

Illustrator FLAVIA BACARELLA, Stewart's wife, is an artist who teaches painting and drawing at Lehman College of the City University of New York.